SpringerBriefs in Philosophy

SpringerBriefs present concise summaries of cutting-edge research and practical applications across a wide spectrum of fields. Featuring compact volumes of 50 to 125 pages, the series covers a range of content from professional to academic. Typical topics might include:

- A timely report of state-of-the art analytical techniques
- A bridge between new research results, as published in journal articles, and a contextual literature review
- A snapshot of a hot or emerging topic
- An in-depth case study or clinical example
- A presentation of core concepts that students must understand in order to make independent contributions

SpringerBriefs in Philosophy cover a broad range of philosophical fields including: Philosophy of Science, Logic, Non-Western Thinking and Western Philosophy. We also consider biographies, full or partial, of key thinkers and pioneers.

SpringerBriefs are characterized by fast, global electronic dissemination, standard publishing contracts, standardized manuscript preparation and formatting guidelines, and expedited production schedules. Both solicited and unsolicited manuscripts are considered for publication in the SpringerBriefs in Philosophy series. Potential authors are warmly invited to complete and submit the Briefs Author Proposal form. All projects will be submitted to editorial review by external advisors.

SpringerBriefs are characterized by expedited production schedules with the aim for publication 8 to 12 weeks after acceptance and fast, global electronic dissemination through our online platform SpringerLink. The standard concise author contracts guarantee that

- an individual ISBN is assigned to each manuscript
- each manuscript is copyrighted in the name of the author
- the author retains the right to post the pre-publication version on his/her website or that of his/her institution.

Michael Newall

Impossible Colours: Colour Experience, Vision Science, and the Philosophy of Perception

Springer

Michael Newall
University of Adelaide
Adelaide, Australia

ISSN 2211-4548 ISSN 2211-4556 (electronic)
SpringerBriefs in Philosophy
ISBN 978-3-032-15136-0 ISBN 978-3-032-15137-7 (eBook)
https://doi.org/10.1007/978-3-032-15137-7

This Springer imprint is published by the registered company Springer Nature Switzerland AG
The registered company address is: Gewerbestrasse 11, 6330 Cham, Switzerland

My love wears forbidden colours

—David Sylvian and Ryuichi Sakamoto,
Forbidden Colours, 1983

To Roman and Zazie

Acknowledgements For generous advice and assistance on various aspects of this book as it was finding its shape, I am grateful to Vincent A. Billock, Eli Brenner, Alan Lee and Paul Nedelisky. I am also thankful to an anonymous referee for Springer for sympathetic and critical feedback on the penultimate draft of the book.

I was invited by Michael Kutschbach and Ash Tower to present a paper on the colour brown to their visual culture research group at the University of South Australia in 2021, and the response of that audience encouraged me to work further on that rather niche topic, which eventually became the germ of Chap. 3. Parts of Chap. 4 were presented to the London Aesthetics Forum, University of London, in 2017, the conference Depiction, Pictorial Experience, and Vision Science held at the University of Glasgow in 2018, the Scottish Aesthetics Forum, University of Edinburgh, in 2018, and the Australasian Society for Philosophy and Psychology 2022 annual conference. I am grateful to the organizers and audiences of those events for encouragement and critical feedback. Colleagues and graduate students at the University of Kent's Aesthetics Research Centre also provided valued feedback, and I am particularly thankful to the University of Kent students who took part in the experiment I describe in Chap. 4.

Permission to reproduce images was kindly given by Youping Xiao, Shiming Tang, Norman Duke, James Elkins, and Shaun Ryan. I thank them for that and am indebted to them for the inspiration that their images provided. Thanks also to my brother, John Newall, for assistance with producing some of the other images. Finally, I owe a debt of gratitude to Eleen M. Deprez for her constant support and encouragement, intellectually and emotionally.

Competing Interests The author has no competing interests to declare that are relevant to the content of this manuscript.

Contents

List of Figures

Chapter 1
Introduction

Abstract This brief introductory chapter presents the topic of the book, impossible colours, and its aims, to give new accounts of colour experience and the mechanisms of the visual system that shape it. Impossible colours are forbidden by opponent processing, the prevailing account of how colour is encoded in the visual system, first proposed by Ewald Hering. Despite this, we can and do experience such colours. The book explores two kinds of these colours: (i) elementary colours beyond those proposed by Hering, and (ii) reddish greens and yellowish blues, which combine elementary colours in illicit ways on Hering's account. Our experience of impossible colours requires us to rethink the accepted phenomenology of colour, which is shaped by the mechanisms proposed by Hering, and those mechanisms themselves. This chapter lays out this challenge and describes how the book will respond to it.

Keywords Colour · Colour experience · Impossible colours · Reddish green · Unique colours

This book is about impossible colours. Impossible colours, in science and philosophy, earn their name because they are forbidden by opponent processing, the widely held account of how the visual system encodes colour. Like other writers, I use this name, but I do not believe that these colours are impossible in any logical or physical sense. In fact, I argue that we can and do see such colours—and that one implication of this is that we must reject the accepted understanding how the visual system encodes colour. The book investigates two different kinds of impossible colours. First, there are reddish greens and yellowish blues. They combine hues in ways that contradict opponent processing, and are widely known as impossible or forbidden colours in the literature on colour. Although they are often said to be impossible to see or imagine, we will see a range of evidence that shows this is false, including colour figures that might allow the reader to see these colours themselves. The second kind of impossible colour is entirely familiar, indeed mundane. It includes colours such as brown, purple and orange. I argue that the phenomenology of these colours is wrongly understood, and that understood correctly, they too contradict opponent processing. Brown, for example, is usually said to appear as a blackish or blackened yellow or orange colour.

M. Newall, *Impossible Colours: Colour Experience, Vision Science, and the Philosophy of Perception*, SpringerBriefs in Philosophy,
https://doi.org/10.1007/978-3-032-15137-7_1

This, I argue, is wrong: brown is not a composite colour, but is in fact elementary to our colour experience. However, understanding brown and these other colours as elementary is also forbidden by opponent processing. So these colours, too, can be called impossible colours, although I will largely reserve the name for reddish greens and yellowish blues in what follows. My study provides a new account of colour experience, to which these two kinds of colours provide the keys, and explores its deep implications. For science, it calls into question the opponent process theory. That theory is also increasingly recognized as inadequate or incomplete by vision scientists, and I show how recent discoveries in vision science allow us to understand how the colours I discuss could be encoded in the visual system. The colours also unsettle philosophical theories of colour and perception, providing new challenges to address. I dwell especially on how they give new perspectives on long-contested questions. Are colours characteristics of the external world, which our visual systems grant us direct experiential access to? Or are colours generated in some way by the visual system, projected onto the external world—and a means by which we come to know it?

* * *

It is often said that humans with normal colour vision are capable of discerning millions of different colours—"at least 10 million", according to one source (Judd and Wyszecki 1975, 345). That is a vast number, but it will also be apparent that the great majority of these colours can be described in terms of other colours. Blue, for instance, can be perceived in darker or lighter shades, and all these shades of blue may appear blended with other hues—for instance, there may be greenish blues or purplish blues. Such shades and blends count for most of the colours that we can experience. These shades and blends are all mixtures. By *mixture*, I do not mean physical mixtures: subtractive colour mixture, as seen in the mixture of paints or overlaid coloured films, nor additive colour mixture, as seen in the mixture of coloured lights. Rather, I mean a "phenomenal" mixture—the visual experience of different colours appearing present in precisely the same part of a single surface, light or substance. A phenomenal mixture of colours *a* and *b* is just a colour, *c*, that is simultaneously experienced as both *a* and *b*, and only as *a* and *b*. That is, *a* and *b* together exhaust the phenomenal character of *c*. Between blue and green in the spectrum, for instance, there are a great many distinct combinations of these colours, the colour becoming bluer in one direction and greener in the other. These blues are to varying degrees greenish and bluish—and that is all there is to them in phenomenal terms. The spectral colours are usually thought to combine in phenomenal mixtures with black, white and grey to create all the other colours we can experience (Op. cit., 255). But some colours cannot be understood as combinations. These are *elementary* colours—or unique colours, as they are sometimes called—the phenomenological elements of our colour experience.[1] Blue and green, for example, seem to be elementary colours. Blue and

[1] The scientific and philosophical literature often treats the terms "elementary colour" and "unique colour" interchangeably. However, in some discussions, "unique" colours describe elementary colours in their unmixed form.

green mix with other colours, but it does not seem possible to analyze blue and green into further colours.[2] This study seeks to answer two questions. First, which colours are elementary? Second, in what combinations can the elementary colours appear together, creating the myriad of colours we experience?

In taking up these questions, this study develops an alternative phenomenology of colour to what I will call the orthodox theory, that of Ewald Hering (1834–1918), which is currently accepted as standard in science and philosophy. Hering's approach is shaped by two principles. The first holds that those with normal colour vision see just six elementary colours: red, yellow, green, blue, black and white. All the other colours, on this account, are phenomenal mixtures of these. The second principle holds that certain elementary colours never appear together in phenomenal mixtures. In particular, there are no reddish greens and yellowish blues. Together these principles shape familiar three-dimensional colour spaces ranging from Runge's *Farben-Kugel*, or colour sphere (Runge 1810), to currently used models such as the Munsell colour space (Judd and Wyszecki 1975, 258–266), and the NCS colour system (Hård et al. 1996).

Why should we be interested in the phenomenology of colour, and what hangs on questioning the orthodox theory? Prior to the nineteenth century, focus on the experience of colour was unsystematic, with scientists, philosophers and artists concerned largely with colours of physical substances, and coloured lights (Kemp 1992). That began to change with Chevreul's (1839) work on subjective contrast effects, and with the experimental work of Helmholtz (1924–25). Hering (1964) was the first to systematically consider colour experience as a topic disconnected from its stimulus conditions, and, as I have said, his account remains standard.[3] Perhaps one reason why Hering's account has been so little questioned is that it is closely linked to the opponent process theory. Hering proposed the opponent process theory on the basis of his phenomenology, and in the 1950s mechanisms of opponency were discovered (Hurvich and Jameson 1957). In vision science, Hering's account of colour experience and opponent processing have stood as mutually reinforcing theories since then. His account of colour experience is understood as revealing the physiology of opponent processing, and opponent processing in turn is understood as shaping and constraining colour experience. Any new account of colour experience will thus have implications for thinking about the physical processing and encoding of colour information in the visual system. An aim of this book is therefore to explore the implications of a new account of colour experience for thinking about the visual system.

---

[2] I put aside the possibility that blue in fact divides into two different elementary hues, a light blue and dark blue. I take this up later, in Chap. 3.

[3] In the visual arts, Hering's account of colour experience and the opponent process theory are little known, as artists lost interest in new theoretical developments about colour perception after the nineteenth century. That may seem surprising, given twentieth-century art's concern with colour. One justification is found in the approach of influential modernist artist and educator Josef Albers, who wrote that his teaching "does not begin with optics and physiology of visual perception, nor with any presentation of the physics of light and wave length. …[N]o color system in itself can develop one's sensitivity for color. (Albers 1975, 1–2).

In philosophy, Hering's views arrived some time later, introduced in a systematic way in 1988 in C. L. Hardin's *Color for philosophers* (Hardin 1993). A key question raised by philosophers about colour is whether it is objective, a physical quality of surfaces; subjective, somehow produced by the visual system of the viewing subject; or some other kind of property. The nature of colour experience could be decisive in resolving that debate, for broadly speaking, if colour experience tracks physical qualities of surfaces, that tends to support objectivism, and if it better tracks patterns of activation in our visual system, that tends to support subjectivism. That debate also has implications for broader philosophical discussions. Objectivism implies that we have a transparent, direct perceptual access to the world around us—the world, then, would be largely how it appears to us (see Reid (1997) for a classic formulation, and Martin (2002) for a recent interpretation). Subjectivism implies that our perceptual access to the world is indirect (see Locke (1979) for a traditional source), and that the world as it appears is a construction of the mind, a "grand illusion" (Noë 2002), or "controlled hallucination" (Seth 2021).[4] Hardin interpreted this new data derived from Hering as supporting subjectivism (Hardin 1993, 59–112). Others were galvanized into new defenses of objectivism (e.g., Byrne and Hilbert 2003). Similarly, revisions to Hering's phenomenology of colour hold the potential for new challenges and insights for philosophy of colour. Thus, a final aim of this book is to outline the implications that my account of colour experience has for philosophy of colour.

I develop my arguments as follows. Chap. 2 presents Hering's account of colour experience—that is to say, the orthodox theory—and the pair of principles that shape it. The following two chapters present evidence that both these principles are wrong. In Chap. 3, we shall find that there are more elementary colours than Hering recognizes. In Chap. 4, we will see there are credible reports of reddish greens and yellowish blues. Together, these results produce a much richer phenomenology of colour than that of the orthodox view, and, I believe, more accurately reflect the character of our own colour experience. As I have said, my account of colour experience will meet a challenge in the opponent process theory, which prohibits the elementary colours beyond those proposed by Hering, along with greenish reds and yellowish blues. Chap. 5 responds by showing that the phenomenology of colour I outline finds support in recent scientific research on the neurophysiology of vision that suggests the possibility of a form of colour coding different to opponent processing. Chap. 6 gives an overall picture of the phenomenology of colour suggested by these results, and explores the implications for accounts of colour space. Lastly, Chap. 7 explores the implications for philosophy of colour, finding that objectivism faces difficult challenges and must abandon some of the intuitions that motivate it if it is to accommodate my results.

Beyond reddish greens and yellowish blues there are other kinds of colours, more or less well-understood, that are described as lying outside the bounds of ordinary human experience and the ordinary operation of the human visual system.

[4] While Seth argues that the visual world is a "controlled hallucination", Noë rejects the idea that the visual world is a "grand illusion".

Macpherson (2021) gives a helpful survey of these. They are not the focus of my study, but they are often the subject of fascination, and so I footnote some information about them, describing how they relate to the topic of my study where they do so.[5]

References

Albers J (1975) Interaction of color. Revised edn. Yale University Press, New Haven, Conn

Broackes J (2010) What do the color-blind see? In: Cohen J, Matthen M (eds) Color ontology and color science. MIT Press, Cambridge, Mass, pp 291–405

[5] Chimerical colours (Churchland 2005) are colours that appear brighter or darker than they can in ordinary vision, due to the exploitation of contrast effects. These include "stygian" colours, dark colours which seem more saturated than such colours can normally appear, "self-luminous" colours, which appear to glow, and "hyperbolic" colours, which seem more intensely saturated than colours can normally appear. Chimerical colours are sometimes described as impossible, since they are outside the realm of ordinary colour experience. I touch on stygian colours in Chap. 3, and include a stimulus for inducing a stygian colour there. Synaesthetic colours are products of synaesthesia, in which the stimulation of one sense or perceptual capacity, produces activity in another—in this case, in colour perception. In some cases, synaesthetic colours are reported as being different in kind to colours experienced in response to normal visual stimuli. These synaesthetic colours have been called "weird" (Ramachandran and Hubbard 2001a, 982), "alien" and "Martian" (Ramachandran and Hubbard 2001b, 26–27). Lastly, there are colours perceived by those with different kinds of colour receptors in their retinas, in comparison to the ordinary three that most of us have. Many animals have such different colour perception, and as one would expect, there is a lot of mystery around their colour experience. For instance, many birds have a fourth photoreceptor that allows them to respond to light in the ultraviolet range (Chen et al. 1984). They thus have tetrachromatic vision, in contrast to the trichromacy of ordinary human vision. Mantis shrimp have twelve kinds of colour receptors, although, curiously, this does not seem to give them excellent colour vision (Thoen et al. 2014). Some people also have a fourth kind of colour receptor, usually receptive to stimuli in the yellow-orange part of the spectrum (Jordan et al. 2010; Jordan and Mollon 2019). This form of tetrachromacy requires one to have two X chromosomes—which is to say, most human tetrachromats are cisgender women. Human tetrachromats can have an enhanced acuity in their perception, and I have argued that this likely allows them to see colours radically different from those that ordinary, trichromat viewers perceive (Newall 2025). There is also a relatively simple way to artificially acquire tetrachromat vision functionally similar to that of natural tetrachromats. This involves the use of optical filters, ideally fitted into glasses (Gundlach et al. 2018). One of the filters blocks most of the light coming from the red part of the spectrum so that the red-receptive cells in one eye are effectively turned into "orange" receptors. Binocular vision will then combine input from the "orange" receptors with input from the "red" receptors in the other eye, together with the input from the other, ordinary, colour receptors, giving tetrachromat input in the centre of the field of vision. It takes a little while to adapt to the filters, and they need to be used in the right visual environment to get their full effect (well-lit and showing subtle distinctions of colour in the red/orange/yellow part of the spectrum). The colours one will experience are not novel colours that I think natural human tetrachromats experience. Rather, it seems that they are impossible colours, such as reddish greens and yellowish blues (an observation made by Lachner, n.d.). Sometimes, those with colour-blindness have been described as seeing not a reduced range of colours compared those with ordinary vision (a well-established position found in Viénot et al. 1995; Hardin 1993, 145–46), but rather as experiencing different kinds of colours. A sophisticated account on these lines, drawing in part on his own experience, is given by Justin Broackes (2010).

Byrne A, Hilbert DR (2003) Color realism and color science. Behav Brain Sci 26:3–64
Chen DM, Collins JS, Goldsmith TH (1984) The ultraviolet receptor of bird retinas. Science 225:337–340
Chevreul M-E (1839) De la loi du contraste simultané des couleurs. Pitois-Levrault, Paris
Churchland PM (2005) Chimerical colors: some phenomenological predictions from cognitive neuroscience. Philos Psychol 18:527–560
Gundlach BS, Frising M, Shahsafi A, Vershbow G, Wan C, Salman J, Rokers B, Lessard L, Kats MA (2018) Design considerations for the enhancement of human color vision by breaking binocular redundancy. Sci Rep 8:1–9
Hård A, Sivik L, Tonnquist G (1996) NCS, Natural color system—From concept to research and applications. Part I Colour Res Appl 21:180–205
Hardin CL (1993) Color for philosophers: unweaving the rainbow, expanded. Hackett, Indianapolis
Helmholtz HV (1924–25) Helmholtz's treatise on physiological optics, JPC Southall (ed) Optical Society of America, Rochester, NY
Hering E (1964) Outlines of a theory of the light sense (trans: Hurvich LM, Jameson D). Harvard University Press, Cambridge, Mass
Hurvich LM, Jameson D (1957) An opponent-process theory of color vision. Psychol Rev 64:384–404
Jordan G, Deeb SS, Bosten JM, Mollon JD (2010) The dimensionality of color vision in carriers of anomalous trichromacy. J Vis 10:1–19
Jordan G, Mollon JD (2019) Tetrachromacy: the mysterious case of extra-ordinary color vision. Curr Opin Behav Sci 30:130–134
Judd DB, Wyszecki G (1975) Color in business, science, and industry, 3rd edn. John Wiley & Sons, New York
Kemp M (1992) The science of art: optical themes in western art from Brunelleschi to Seurat. Yale University Press, New Haven, Conn
Lachner K-R. n.d. True-red non-retinal (chromatically-less-redundant) moderately functional tetrachromacy. https://www.color-in-color.info/tetrachromacy_1/non-retinal-tetrachromacy. Accessed 5 Apr 2025
Locke J (1979) An essay concerning human understanding. Clarendon Press, Oxford
Macpherson F (2021) Novel colour experiences and their implications. In: Brown DH, Macpherson F (eds) The Routledge handbook of philosophy of colour. Routledge, Abingdon, pp 175–209
Martin MGF (2002) The transparency of experience. Mind Lang 17:376–425
Newall M (2025) What is it like to be a tetrachromat? Philos Mind Sci 6. https://doi.org/10.33735/phimisci.2025.11645
Noë A (ed) (2002) Is the visual world a grand illusion? Imprint Academic, Exeter
Ramachandran VS, Hubbard EM (2001a) Psychophysical investigations into the neural basis of synaesthesia. Proc R Soc Lond B 268:979–983
Ramachandran VS, Hubbard EM (2001b) Synaesthesia—a window into perception, thought and language. J Conscious Stud 8:3–34
Reid T (1997) An inquiry into the human mind on the principles of common sense. Brookes DR (ed), Edinburgh University Press, Edinburgh
Runge PO (1810) Farben-Kugel oder Construction des Verhältnisses aller Mischungen der Farben zu einander, und ihrer vollstëndigen Affinität, mit angehängtem Versuch einer Albeitung der Harmonie in den Zusammenstellungen der Farben. Friedrich Perthes, Hamburg
Seth A (2021) Being you: a new science of consciousness. Faber & Faber, London
Thoen HH, How MJ, Chiou T-H, Marshall J (2014) A different form of color vision in mantis shrimp. Science 343:411–413
Viénot F, Brettel H, Ott L, Ben M'Barek A, Mollon JD (1995) What do colour-blind people see? Nature 376:127–128

Chapter 2
The Orthodox View of Colour Experience

Abstract This chapter gives an account of Ewald Hering's description of colour experience, which has become the orthodox view in vision science and philosophy. It also examines how this shapes Hering's account of the encoding of color information within the visual system—the opponent processing theory—and indicates how the theory has begun to come under pressure within vision science.

Keywords Colour experience · Ewald Hering · Opponent processing · Unique colours

What I call the orthodox view of colour experience, developed by Ewald Hering (1964) is shaped by two principles. The first of these holds that those of us with normal vision experience six elementary colours. These are achromatic—black and white—and chromatic—red, yellow, blue, and green. Hering understands every other colour we can see as a mixture—a phenomenal mixture—of two or more of these elementary colours. If we attend carefully to any non-elementary colour, we should be able to give an exhaustive description of its phenomenal character in terms of its elementary components (although there is, we shall see, some discussion about just how easy this is in some cases) So, for Hering, orange is a phenomenal mixture of red and yellow—it is simply a colour that appears both reddish and yellowish (Hering 1964, 42). Purple is a reddish-bluish colour. (Ibid.) Brown is usually said to be either a blackish yellow colour (op. cit., 58), or a blackish orange colour (Hardin 1993, 141)—which is to say, a blackish, reddish and yellowish colour, since Hering understands orange as a reddish yellowish colour.

The second principle that shapes the orthodox view holds that there are restrictions on how colours can mix in our experience. In particular, Hering observed, red and green never appear mixed, and nor do yellow and blue. That is to say, there are no reddish greens or yellowish blues. (Hering 1964, 49–50) It is in this sense that red and green, and yellow and blue, are "opponent" colours. Together these principles give familiar colour spaces their structure. This is most clear in a limited colour space, the colour wheel, which arranges the chromatic colours (or hues) so that those that appear more similar to each other are closer to each other. Any broad selection of

M. Newall, *Impossible Colours: Colour Experience, Vision Science, and the Philosophy of Perception*, SpringerBriefs in Philosophy,
https://doi.org/10.1007/978-3-032-15137-7_2

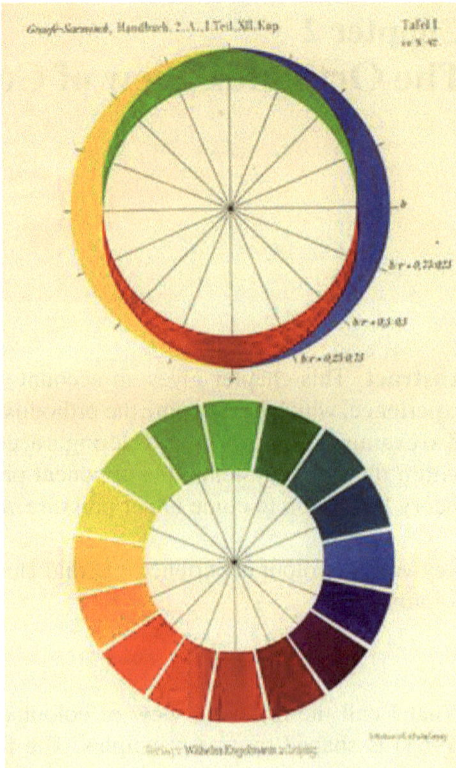

Fig. 2.1 Ewald Hering, Diagram illustrating how different proportions of Hering's elementary hues (top) constitute the hues of the colour wheel (bottom). From Hering (1964, pl. 1)

samples of chromatic colours, arranged according to this rule, will take the form of the circle at the bottom of Fig. 2.1: a colour wheel. The circle at the top of Fig. 2.1 shows Hering's chromatic elementaries—red, yellow, blue and green—illustrating how the colours of the lower circle are (on his account) constituted on experience as varying mixtures of his four chromatic elementaries. But while non-opponent elementaries can appear mixed with one another, the opponent colours cannot—so there is no space for reddish green and yellowish blue on the colour wheel.

The colour wheel shows Hering's chromatic elementaries, and all the mixtures which on his account they can give rise to. On the orthodox view, a complete colour space can be constructed by adding black, shades of grey, and white, together with the various mixtures these can make with the chromatic colours. I will call these orthodox colour spaces, as I will introduce another version of colour space in Chap. 6. The earliest orthodox colour space is found in Philipp Otto Runge's *Farben-Kugel* or colour sphere of 1810 (Fig. 2.2), "the first topologically correct account of the full color gamut", as Koenderink (2008) calls it. Around the sphere's "equator" is the familiar colour wheel, at one "pole" is white, and at the other is black. At the centre of the sphere, on the axis connecting black and white, lies grey, and around this core lie greyish tints of the chromatic colours. Proceeding towards the black

Fig. 2.2 Two views of Philip Otto Runge's *Farben-Kugel* or colour sphere. From Runge (1810, pl. 1)

pole the chromatic colours become increasingly darker, appearing tinted with black, and towards the white pole they become increasingly brighter, appearing tinted with white. Modern colour spaces also have this orthodox colour structure, such as the Natural Color System (NCS), which draws directly on Hering.[1] (Hård et al. 1996) Unlike Runge's *Farben-Kugel*, the NCS colour space is a flattened spindle shape. It is encircled at its widest point by the chromatic colours of the colour wheel, with red, yellow, green and blue—Hering's elementary hues in their pure states—marking four evenly spaced "compass" points.

Hering proposed the opponent process theory to explain these apparent facts about elementary colours and their mixture. The opponent process theory holds that chromatic colours are encoded in the visual system in two opponent channels: one for red and green, and the other for yellow and blue. The cells in each of these channels respond in one way to encode one colour, and in the opposite way to encode the opponent colour. So, the system can encode all the colours of the colour wheel, but cannot encode reddish greens and yellowish blues. A further channel encodes black and white, but it is not an opponent channel, as it allows black and white to mix, creating colours that appear at once blackish and whitish—greys on Hering's account.[2] Together, the three channels encode all the colours normal viewers are capable of seeing for Hering, which is also to say all the colours that appear in the colour spaces in Fig. 2.2. A neural basis for the channels that Hering describes was discovered by Hurvich and Jameson (1957), and it is clear that signals are encoded this way in the optic nerve and into the brain. Hering's spare and elegant phenomenology, together with the physiological discoveries that are taken to explain it, are a powerful coupling. The phenomenology has come to be widely accepted by vision scientists: "All [chromatic] colors can be described in terms of four non-reducible 'unique' hues: red, green, yellow, and blue" (Stoughton and Conway 2008, R698) "Hue can

[1] The Natural Color System shares its name with Hering's essay, 'The natural color system' (Hering 1964, 25–65).

[2] "[E]very gray is at the same time whitish and blackish, now it is more of the one, and now more of the other, or in a special case it may be both to about the same degree" (Hering 1964, 31–32).

be described by four separate sensations of red (R), green (G), yellow (Y), and blue (B)" (Abramov and Gordon 2005, 2143). "There exist four colours, the *Urfarben* of Hering, that appear phenomenologically unmixed" (Mollon and Jordan 1997, 38). In philosophy, too, where Hardin first introduced these ideas, they have become the orthodox way of thinking about colour phenomenology and its underlying physiology (Hardin, 1993), and although many have questioned Hardin's subjectivism about colour, this phenomenology and physiology are widely understood and accepted in philosophy of perception.[3] (Maund 2018).

I will not reject wholesale Hering's physiological account of colour processing. Colours are encoded as Hering describes as they leave the eye and into the visual pathways of the brain—as Hurvich and Jameson discovered—and this shapes our phenomenology in important ways. But it is also recognized among colour scientists that Hering's account of colour experience does not wholly correspond to the coding that occurs in opponent process pathways:

> [M]ost colour scientists are agreed that the chromatically opponent cells of the early visual system … do not correspond colorimetrically to red-green and yellow-blue processes. (Mollon and Jordan 1997, 382)
>
> [O]pponent processing cells at this early stage are "not the correlate for seeing a particular hue quality, say unique [i.e. elementary] red. This latter correlate, if it exists as a separable entity, must be associated with yet unidentified, higher-level neural activities. (Valberg 2001, 1645)

So, it appears that further coding of colours must occur in some other form. This will be key to my account, as there needs to be a prospect of a non-opponent coding of colour in the visual system if my phenomenology is to be consistent with visual science. That is the possibility I pursue in Chap. 5.

References

Abramov I, Gordon J (2005) Seeing unique hues. J Opt Soc Am A 22:2143–2153

Hård A, Sivik L, Tonnquist G (1996) NCS, Natural color system—from concept to research and applications. Part I Colour Res Appl 21:180–205

Hardin CL (1993) Color for philosophers: unweaving the rainbow, expanded. Hackett, Indianapolis

Hering E (1964) Outlines of a theory of the light sense (trans: Hurvich LM, Jameson D). Harvard University Press, Cambridge, Mass

Hurvich LM, Jameson D (1957) An opponent-process theory of color vision. Psychol Rev 64:384–404

Koenderink J (2008) Review of R. G. Kuehni, Philipp Otto Runge's color sphere: a translation, with related materials and an essay. Color Res Appl 33:512–513

Maund B (2018) Color science—some complexities. supplement to color. In: Zalta EN (ed) Stanford encyclopedia of philosophy (Spring 2022 edn). https://plato.stanford.edu/entries/color/color-science-complexities.html

[3] Maund observes that Hardin (1993) "was significant in bringing to the attention of philosophers the enormous amount of progress that has been made … particularly in the realm of color vision" (Maund 2018).

Mollon JD, Jordan G (1997) On the nature of unique hues. In: Dickinson C, Murray I, Carden D (eds) John Dalton's color vision legacy. Taylor & Francis, London, pp 381–392
Runge PO (1810) Farben-Kugel oder Construction des Verhältnisses aller Mischungen der Farben zu einander, und ihrer vollstëndigen Affinität, mit angehängtem Versuch einer Albeitung der Harmonie in den Zusammenstellungen der Farben. Friedrich Perthes, Hamburg
Stoughton CM, Conway BR (2008) Neural basis for unique hues. Curr Biol 18:R698–R699
Valberg A (2001) Unique hues: an old problem for a new generation. Vision Res 41:1645–1657

Chapter 3
Elementary Colours Beyond Red, Yellow, Green, Blue, Black and White

Abstract According to Ewald Hering, there are six elementary colours—red, yellow, green, blue, black and white. Here I argue that a careful examination of experience—one's own and reports of others—will show that there are more elementary colours than Hering recognizes. In particular, there are good reasons to think that those of us with normal colour vision tend to experience also orange, purple and brown as elementary. I also suggest that some may experience a few further colours as elementary, including grey and pink, and light and dark blue.

Keywords Colour experience · Unique colours · Stygian colours · Colour terms

Ewald Hering (1964) claims that there are just six elementary colours: red, yellow, green, blue, black and white. This chapter argues that normal viewers also perceive other colours as elementary—in particular, orange, purple and brown—and there may be further elementary colours beyond these.

A good part of the case I build relies on reports of individuals' experiences. I also describe ways the reader can draw on the evidence of their own colour experiences to test my proposal. One might be concerned that this approach cannot do enough to make any alternative to Hering's theory compelling. So let me start by addressing some immediate questions the reader may have about method. Hering never conducted experiments on which colours are elementary—he relied on the reader's experiences matching his own descriptions to make his phenomenology plausible. Indeed, few such experiments have been conducted, and these are not especially helpful in resolving the question which colours are elementary.[1] Hering and his successors largely take his phenomenology as self-evident, rather than providing experimental evidence for it. So, my method to investigating colour experience is in no way inferior to Hering's. One might reasonably worry that the reports I draw on only justify conclusions about how *some* people experience colour. As will become

---

[1] Lindsey et al. (2020) find that Hering's elementaries are not always cross-culturally applicable, but they begin from the assumption that they correctly apply among Western observers. Fuld et al. (1983) use empirical methods to argue that brown could be an elementary character, initiating an inclusive back and forth exchange on the question. See Newall (2022) for a discussion of this.

M. Newall, *Impossible Colours: Colour Experience, Vision Science, and the Philosophy of Perception*, SpringerBriefs in Philosophy,
https://doi.org/10.1007/978-3-032-15137-7_3

apparent, I am open to the idea that not all of those of us with normal vision will experience the same colours as elementary, although my sense is that such differences are fairly limited. Still, even if it turned out that I was wrong about this, and some people experienced orange, purple and brown as elementary while others did not, that limited result would still be of interest, and would still show that the opponent theory cannot hold—at least, not as a general theory. What if the reports I draw on are mistaken about the experiences they describe? The accounts I draw on all come from those with scientific or philosophical expertise in colour, and they are, I think, well-apprised of the fact that their observations are impossible on Hering's account. Of these, only one argues the case for an alternative to the opponent theory (Conway et al. 2023). For the others their observations count as awkward facts, which they mention precisely because they find them so obtrusive in the context of Hering's theory.

The ideas in this chapter were initially prompted by poppies growing in my front yard. (Fig. 3.1) Around the outer edges of the flowers' petals, they seemed to me an orange-yellow, and towards the centre of the blooms they appeared increasingly orange. Yet looking at the change in colour from the flowers' edges to their centres I could not persuade myself that I was really seeing a yellowish colour that was becoming increasingly reddish. That is to say, my perception failed to accord with the orthodox view that orange is simply a reddish yellowish colour. (Orange, as Hardin puts it, "could not retain its identity were it not reddish and yellowish" (Hardin 1993, 126). Towards the centre of the blooms, the petals became more orange, but they did not appear to become more red—they just didn't look reddish to me in any part, no matter how much I reflected on it.

Our perception of colours in natural scenes is often influenced by complex factors such as contrast effects and associations. So, judging colours outside such scenes can sometimes yield different results. Looking at the squares of yellow, orange and red in Fig. 3.2, however, I find that the orange square still resists appearing simply a mixture of red and yellow, examples of which appear on each side for comparison.[2] Some oranges are reddish or yellowish (that shown below might seem to be tinted

---

[2] The colours of these squares are not equiluminant—that is, the yellow is a light colour, the orange is darker, and the red is darker again. I have given saturated examples of the three colours so that each hue can be considered *qua* hue. If the colours are adjusted so they are equiluminant, the yellow becomes so dark that its recognizably yellow colour risks being lost (it becomes olive to my eyes). However, the lack of equiluminance introduces a risk that our assessment of the colours is unwittingly influenced by their differences in luminance rather than hue. That is, we might worry that we are wrongly led to reject the idea that orange is a yellowish reddish colour because it differs in luminance from yellow and red. I would respond by pointing out that we have no trouble recognizing some colours as being comprised of others, even when the saturated form of their components have rather different luminances. For instance, yellow and green have different luminances in their saturated forms, but a phenomenal mixture of the two, a yellowish green, is recognizable as just that—as sharing its yellowish and greenish components with the saturated pure versions of yellow and green. So I see no reason why we should not be able to do the same in the case of orange if it really were experienced as a reddish yellowish colour. Note too that this concern about a lack of equiluminance cannot be raised about the red, purple and blue squares in Fig. 3.4, since they are very close to equiluminant.

Fig. 3.1 Californian poppies. *Photo* the author

Fig. 3.2 Squares of yellow, orange and red

with one of these colours to some viewers), but to my eye at least, there is always a quality of colour present in orange that is *neither* reddish nor yellowish, nor does it appear to be mixed—phenomenally comprised—of any other colours. That is to say, orange seems to me an elementary colour.

I am not the only one to make such observations. In particular, two philosophers have made the same observation. Wayne Wright observes:

> I can find orange … chips in the *Munsell Book of Color* that appear to me every bit as basic in hue as any unique Hering primary; I have normal color vision by the usual tests and was born and raised in an English-speaking, Western industrialized nation. I do not detect any

> redness or yellowness … in them, but find only certain similarities to red and yellow … in the same way that I find similarities between blue and green (Wright 2019, 1518).[3]

Radek Ocelák makes the point in this way:

> Orange can be said to perceptually contain red and yellow, yet literally there is as little of red in orange as there is, literally, of luck in odd numbers. The perceptual reality as regards red, yellow and orange is not that orange in any literal manner consists of the former two, but that orange lies between red and yellow in the perceptual space of colors …. (Ocelák 2015, 517)

Scientists too have made this observation. R. M. Boynton notes:

> Orange … contains (for some people at least) something that is rather different from just a blend of red and yellow …. (Boynton 1975, 316)

Neuroscientist Bevil Conway and his colleagues also have this in mind when they ask:

> Is the saturated peel of an orange really "reddish yellow"? Or is orange unique? (Conway et al. 2023, 795)

Here is a further way to assess the elementary or non-elementary character of orange. Consider the colour wheel in Fig. 3.3, which gradates smoothly through all the chromatic colours. Look at that segment in which red gradates to yellow. Does it appear to gradate directly from red to yellow? If it does, that would show that the orange we see there is not an elementary colour, but in fact a mixture of red and yellow. Or does red gradate through an orange colour that cannot be described as simply a blend of red and yellow? If so, then it is further evidence that the orange we see there is an elementary colour.

I have said already that the accounts we have of what the experience of orange is like could support a more qualified conclusion: that some of us experience orange as elementary, and others experience it as a reddish yellowish colour. Similarly, I remain open to the possibilities that different viewers may experience the orange in the figures above as either elementary or as a reddish yellowish colour, and that a single viewer may experience some of these oranges one way, and some the other. But if one or both these possibilities turn out to be true, it remains the case that Hering's phenomenology, and the opponent process theory that he takes to explain, cannot hold at a general level.

Similar questions can be put of purple, and they suggest a similar conclusion, that purple is also experienced as an elementary colour, by at least some normally-sighted individuals, at least some of the time. Consider squares of red, purple and blue in Fig. 3.4. Does purple seem a combination of those colours to each side of it? Do we recognize blueness and redness in the square of purple, and find nothing, phenomenally speaking, left over? Or does the purple square seem to incorporate a

[3] Wright concludes that orange, and also, for similar reasons, purple, are no less elementary to colour experience than Hering's elementaries. From there he takes a different path to me, arguing against the idea that any colours are unequivocally elementary or "unique". Rather, they are only unique relative to various measures—and these measures deliver equivocal results (Wright 2019).

Fig. 3.3 Colour wheel with smooth gradations between colours

Fig. 3.4 Squares of red, purple and blue

colour that is not exhausted by removing any red or blue components one may see in it? If one answers yes to this last question, it implies that the purple we see there is an elementary colour. Turning again to the colour wheel in Fig. 3.3, one may ask, is the purple that appears in between red and blue no more than a bluish reddish colour, or does it gradate through another colour, one that cannot be described in these terms? Here, too, I suggest, the purple we see appears to us as an elementary colour. Again, I am not the only one to describe the experience of purple as elementary. Boynton (1975, 316), Wright (2019, 1518) and Conway et al. (2023, 795) make much the same observation.

If what I have said so far is persuasive, one may still wonder why we have so readily accepted that orange is simply a reddish yellowish colour, and purple a reddish bluish colour? One likely explanation is a tendency to confuse phenomenal mixture with the physical mixture of colours with which most of us are familiar since childhood: paint mixture. Mixing red and yellow paint makes orange paint, and mixing red and blue paint makes purple paint. Here is a striking piece of evidence that suggests that paint mixture can lead us astray about phenomenal colour mixture. Brentano (1907) erroneously believed green was a phenomenal mixture of blue and yellow, his "chief justification" being "the fact that the best colour observers (painters) see blue and yellow in green" (Watt 1908, 128). That is surely due to painters' awareness that the mixture of yellow and blue paint makes green.[4] Brentano's view came to be rejected as Hering's view—that green was an elementary colour—came to dominate vision science.

Now to brown.[5] It is usually said that brown is "a kind of darkened yellow" (Westphal 1982, 420) or, a "blackened orange" (Hardin 1993, 141). That is to say, it is experienced as a mixture of black and either yellow or orange (i.e., reddish yellow on Hering's account). But it is also often observed that brown doesn't really look that way. C. A. Padgham and John Saunders observe, "[w]e can … describe … [browns] as low luminance orange-yellows but they certainly do not look like this" (Padgham and Saunders 1975, 143). Hardin notes that, "the perceptual difference between 'standard' yellow or orange and their blackened counterparts is striking" (Hardin 1993, 141). Mohan Matthen says, "brown and orange seem like different colours, and the sameness of their components is not evident" (Matthen 2021, 166). And comparing brown to dark versions of the chromatic colours (dark red, dark green, and so on) Boynton observes: "[o]f these, brown is certainly the most surprising, because it ceases almost entirely to resemble the original bright color" (Boynton 1975, 316).[6]

One can assess these claims by taking Boynton's approach. Consider the HSL (Hue Saturation Lightness) diagram (Fig. 3.5). Along its horizontal axis it shows the chromatic colours, grading to white at the top and black at the bottom.[7] Trace along the bottom where the chromatic colours fade to black. There is a dark red, dark blue, dark green, and so on. But where one might expect to see a dark orange and dark yellow, there is a smudge of another dark colour: brown. At least, that is how Boynton sees it. Here is a second test. Trace a line downwards from the orange and yellow areas. Where the other chromatic colours fade directly into black via dark versions of themselves, orange (especially) and (some may also say) yellow do not fade directly to black, but fade first to brown. That is to say, the smudge of brown on the HSL diagram seems not to be experienced as a blackish version of orange or

[4] I draw this account of Brentano (1907), which remains untranslated in English, from Watt (1908). On Brentano's view of green, see also Boring (1942, 151). Further evidence against green being simply a yellowish bluish colour can be drawn from Chap. 4. The examples of yellowish blue discussed there do not appear at all greenish.

[5] The arguments that I give here for the elementary character of brown are much more extensively explored in Newall (2022)—although the argument, below, using stygian colours is new.

[6] Even Hering pauses over his phenomenology as it applies to brown: "the distinguishing feature of brown, namely, its blackness, never shows up clearly as an independent quality added to the yellow hue" (Hering 1964, 58).

[7] The diagram can be thought of as providing a map of the colours on the surface of Runge's colour sphere (the map would then be something like a Mercator projection of Runge's sphere) or of colour solids such as the NCS colour space. It does not include the greys, or desaturated version of these colours, that lie within these colour spaces.

Fig. 3.5 HSL (Hue Saturation Lightness) diagram. A brown 'smudge' appears where columns of orange and yellow meet the vertical black band

yellow. Rather, it appears as an elementary colour, one that is not a combination of other colours.[8]

Here is another approach to demonstrating brown's status as an elementary colour. If one could be induced to see blackish yellow or blackish orange, and one could see that such colours were *not* brown, that would make a powerful case for brown itself being elementary rather than a combination of Hering's elementaries. So, can such colours be experienced? I believe so. Paul Churchland has developed a method for inducing the experience of what he calls "stygian" colours—intensely black versions of the chromatic colours (Churchland 2005, 544–546). It is possible to produce "stygian" yellows and oranges—that is to say, intensely blackish yellows and oranges—using this method. Moreover, it seems that while they may look brown to some observers, they may instead retain their yellow or orange appearance to others.

Figure 3.6 allows one to generate these colours. Fixate your eyes on the x at the left, then, after a few seconds, fixate on the x at the right. The blue circle's after-image will appear as an orange circle floating over the black surface, and the purplish-blue circle will produce a yellow after-image floating over the black surface. These colour effects, adopting Churchland's terminology, are stygian orange and stygian yellow, respectively. Churchland describes it as "a yellow that is as dark as the darkest possible black, and yet is still not black" (Churchland 2005, 546). He also suggests that some see it as brown: "[y]ou may well judge it to be some kind of unfamiliar *brown*, rather than yellow." (Ibid., original italics) But clearly some do see it as yellow. Philosopher William Sharp describes stygian yellow as "dark-as-black and yet distinctly yellow" (Sharp 2024, 781). Stygian orange, I suggest, will also be described, by at least some,

[8] For arguments for and against the elementary character of brown, see Fuld et al. (1983) and Quinn et al. (1988). I give a critical assessment of these in Newall (2022).

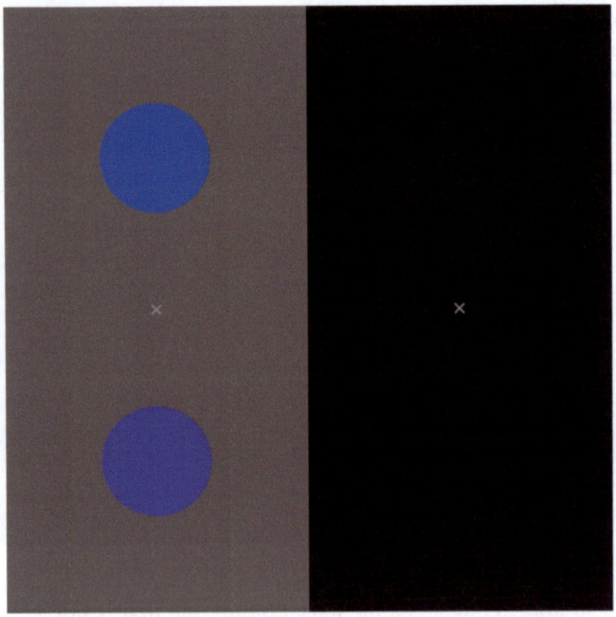

Fig. 3.6 Stimulus for inducing stygian orange (above) and stygian yellow (below). Adapted from Churchland (2005, Fig. 11)

in similar terms: as at once dark-as-black yet distinctly orange, rather than brown. Here, then, is further evidence that brown is an elementary colour. Churchland's method shows us that combining Hering's elementaries, whether black and orange or black and yellow, can produce blackish orange and blackish yellow.[9] It follows that brown cannot be such a mixture, and so must instead be an elementary colour.

One objection that can be raised will say that these stygian colours are experienced not as blackish versions of orange or yellow, but as black grounds seen through transparent areas of orange or yellow. Although Churchland and Sharp do not consider this possibility, I think that it is possible to see these colours in both ways—as a black ground seen through a transparent orange or yellow, and also, by attending carefully to the colour experience, and "blocking out" the transparency effect, as a simple blackish yellow or blackish orange colour, as Churchland and Sharp describe. The fact that these colours *can* be experienced, as Churchland and Sharp attest, is enough

[9] One may ask why Churchland's method can produce a yellow or orange rather than brown. The reason, I suggest, is that the perception of brown requires a contrast effect that does not occur here. In particular, a stimulus must be surrounded by brighter colours if it is to appear brown. ("Brown is always a dark color next to a lighter background" (Feynman et al. 2013, vol. 1, lecture 35, §3). Brown is "dark relative to the brightness of the surrounding field" (Westphal 1982, 419)). But in these instances, the orange and yellow circles do not appear surrounded by a lighter colour—rather they are surrounded by a very dark colour, that is, black. Churchland, I should add, gives a different analysis aiming to explain why some may see stygian yellow as brown (Churchland 2005, 546), but I put that aside here.

to show they exist. The fact that the stimuli used to produce these experiences can also produce other experiences does not change this.

Another objection can be put. This observes that the black in Fig. 3.6 is not that dark. A printed image can only be so dark, and the black seen on a computer screen can be no darker than the screen itself. So couldn't it be that a darker black reliably produces brown? This worry confuses the darkness of the stimulus with the experience of blackness that it produces. Even though the stimulus is not especially dark, *the experience of* blackness that it produces nevertheless can be "as dark as the darkest possible black", as Churchland puts it. This is because the perception of black, like the perception of brown, is produced by a contrast effect which usually relies on the stimulus being surrounded by a significantly brighter colour. Thus, an inky dark black can be produced by a stimulus which is not very dark at all, objectively speaking. This is how the blacks shown on computer screens are produced, including the black of Fig. 3.6 if you are viewing it on-screen.

Let me respond to this by pointing out that these blackish yellows and oranges, produced as a response to these stimuli are quite dark—and appear darker, or more blackish, than some browns, which range from deep chocolate colours to tan and beige. Tan and beige are clearly lighter than the blackish yellows and oranges we are considering. This means that somewhere within the browns that we are capable of seeing, we will be able to find browns with the same darkness, or blackishness, as the dark yellows and oranges we are considering here. So far as these colours can be distinguished, it follows that browns are not darkish or blackish yellows or oranges.[10]

I have argued that orange, purple and brown should be considered elementary colours. It might be objected that we are prone to misjudge these colours as elementary because of the prominence given to them in language. In English, "orange" is widely used, as are "purple" and words such as "violet" and "mauve" which are usually taken to refer to particular tints of purple. "Brown" is also widely used, as are terms indicating tints or shades of brown, such as "beige". Could this influence our judgement about the nature of these colours, encouraging us to attribute to them a phenomenological character they do not in fact have? Such thinking puts the cart before the horse, for it is the prominence of these colours in our phenomenology—which I take to be associated with their elementary character—that explains their prevalence in our language. This is demonstrated by the fact that all languages have names for orange, purple and brown, provided they have a relatively sophisticated colour vocabulary (a notion I will return to shortly) (Berlin and Kay 1969; Kay et al. 2009).

---

[10] Here, one may wonder whether one can find further evidence that orange and purple are elementary colours in a similar way: by producing reddish yellows that do not appear orange, and reddish blues that do not appear purple. This seems to me an appealing idea. The examples of blackish yellows and oranges suggest that generating such reddish yellows and reddish blues could also be possible. However, while I have found some suggestive examples using contrast effects and other colour illusions, I have not found convincing examples or reports of either reddish yellows or reddish blues.

It could also be objected that the conspicuousness of these colours in our phenomenology, and so language, is just a product of their widespread presence (or, one may prefer to say, the widespread presence of their corresponding reflectance profiles) in our visible environment?[11] Brown, especially, is characteristic of the natural environment—in earth, rocks, and leaf litter, for example. Orange and purple are perhaps less widespread, but often mark out features of our environment that are especially salient to us. This objection fastens on to what seems to be a plausible correspondence between language terms and our environment, but I think this correspondence is in fact consistent with orange, purple and brown being elementary colours, and so it is wrongly seen as evidence against my position. Let me explain what I mean. It is plausible that the prominence and salience of colours in the environment shapes our visual systems, whether through evolution, through environmental conditioning or a combination of the two. Their prominence and salience would then play a role in explaining why orange, purple and brown are elementary colours. For example, experiencing a kind of colour as elementary, having it obtrude in experience as a colour distinct from others, would allow subjects to recognize it more quickly and reliably. We tend to identify phenomenal mixtures in a different, longer and more labor-intensive way, by checking individual colours for the components of the mixture. For example, when we search for dark, bluish greens, we have to carefully assess the scene area by area to find where black, blue and green combine. A wider palette of elementaries than Hering describes reduces the need for such tasks to be laboriously performed. Thus, it makes sense that elementariness tracks the prominence and salience of colours in the visible environment.

Consider another objection. One might explain that orange, purple and brown appear quite different from their components in the following way: when their components combine, they acquire a distinctive phenomenal character, a quality that one could call *unity*, which can more or less mask the presence of the components in our experience. Mohan Matthen develops such an argument:

> Though colour has perceptual components, we experience each colour, whether simple or compound as *one* phenomenal whole, not as many phenomenal qualities bound together. … certain shades that people call yellow have the same hue content as certain shades that they call orange and certain others that they call brown. These shades differ only in lightness … [yet] it is very difficult to recognize that these shades are same in hue. This is a case where the components of colour are intimately mixed together: brown and orange seem like different colours, and the sameness of their components is not evident. … If the darkness of these colours were easily separable from their hue, it would be easy to recognize brown as a blackened orange. (Matthen 2021, 166)

He adds that unity, which he sees as a feature of all combinations of elementary colours, only sometimes makes it difficult to identify a colour's components. In other cases, components remain readily identifiable:

[11] A similar position is adopted by Yendrikhovskij (2001), who argues that a comparatively constant frequency of colours in the environment across cultures, rather than facts about our phenomenology, explains the results of Berlin and Kay (1969).

> It is certainly possible to discern hue components in colours of high chromaticness. It is easy, for example, to see … that turquoise is a greenish blue. It is also easy to discern which of two shades of brown is more yellow and which more red. (Matthen 2021, 167)

If Matthen's approach is correct, one would have a tool to resist my claims that there are elementary colours beyond those identified by Hering. The colours that I classify as non-Hering elementaries could be shown to be composed of Hering's elementaries, their actual make-up hidden by the unification of their components. My response to this position is simple—I reject the idea that unity is a feature of colours. On my account, the need for unity does not arise, because the difficulty of discerning components of certain colour mixtures is explained by recognizing that those colour components are not in fact present—for we are dealing with elementary colours which themselves have no components. In the case of brown, we do not need to posit unity to explain our inability to discern brown as constituted of yellow or orange and black, because brown is not in fact constituted of these colours. Rather, brown is as it tends to appear: an elementary colour. That is also to say that I see unity as a kind of a fudge: it is only needed to prop up the phenomenological failings of Hering's approach, and in truth, it only obscures them. With a broader range of elementary colours, the concept of unity is not needed.

If one accepts that there are grounds to think that orange, purple, and brown should be considered elementary colours, it is also natural to ask, just how many elementary colours ordinary observers can see, and what those colours are. I do not propose a definitive response to this question, but I make some points that could contribute toward an answer. While I think there are more elementary colours than Hering recognized, it should also be clear that there is not an unlimited number of elementary colours. This is demonstrated by the fact that mixtures of elementary colours do exist, and make up most of our colour experience. Almost all the individual colours we experience comprise combinations of elementary colours. One way to appreciate this is to look at the colour circle, which presents a selection of these colours. There are a number of elementary hues, but these occupy very small segments of the circle's perimeter, and everything between them appears as combinations of hues.[12] Between yellow and green, for instance, we see hues that are yellowish and greenish to varying degrees. Between orange and yellow we see hues that are orangish and yellowish to varying degrees (here I take it that orange is experienced as an elementary colour), and so on. That is to say, the vast majority of chromatic colours that we see are combinations of elementary hues.

There is also a practical reason why we should expect there to be a relatively limited number of elementary colours. Organisms will benefit from the ability to distinguish quickly and accurately between several categories of colour (once factors such as constancy and illumination effects are also considered), but there is, presumably, a point at which it would be counterproductive to be able to distinguish between a very

[12] One might wonder why the elementary hues do not each occupy a precise point on the colour circle's circumference, rather than a larger part of it. This is so because we cannot make infinite distinctions among the chromatic colours. Each of the hues we can discriminate, including each of the elementary hues, thus occupies a small part, not a point, on the circle's perimeter.

large number of elementary colours, as these will produce categorizations too fine to be useful, and perhaps too complex to process efficiently.

We can look for at least initial guidance on how many elementary colours can be seen by normal observers, and which colours they are, in the World Color Survey (hereafter WCS; its conclusions are provisionally outlined in Berlin and Kay (1969) and the full results presented in Kay et al. (2009)). The WCS studied the words used to name colour in various languages, and found that all use only a small number of "basic colour terms" (BCTs). These are terms that are widely used and understood in a language group, that are not predominately associated with particular objects or substances (such as "ginger" or "turquoise"), and that do not combine other colour terms (as does "blue-green"). In English and many other languages the WCS found exactly eleven BCTs: those that indicate Hering's elementaries, together with brown, orange, purple, pink and grey (Berlin and Kay 1969, 23). In some languages, including Greek, Italian and Russian, more recent research has found that blue is divided into two categories, a light blue and dark blue, each of which has its own BCTs (Androulaki et al. 2006; Paramei 2007; Paramei et al. 2018). So, those languages appear to have twelve BCTs. Kay et al. (2009), stress that BCTs are best thought of as dividing the colour space into separate categories, rather than as designating particular points or areas of colour space. The first colour terms that languages introduce are "black" and "white", which, in the absence of other BCTs in a language, are used to divide colour space into two parts, indicating dark, cool colours, and warm, bright colours. But it seems that at a certain point, after a process of developing or importing different BCTs, a language's vocabulary of these terms will remain static, at either eleven or twelve. Hardin draws attention to a link, noticed by Webster and Kay (2007), between Hering's elementary colours and the WCS's BCTs: the first six BCTs introduced by languages correspond to Hering's six elementary colours. "This suggests", Hardin says, "that the elementary colors are perceptually salient for all language users, making them more likely to be named"—that is, to be given a BCT (Hardin 2013, 3). I am inclined to instead propose that the WCS functions better as evidence for the elementary character of *all* the colours that it identifies as basic. It makes sense that all the elementary colours will be perceptually salient, and so it also makes sense that all and only those colours are named by BCTs. One may, of course, wonder why Hering's elementaries tend to be more perceptually salient than some of those I have proposed.[13] But I think we can find an explanation for this that allows for the prominence of these colours in experience, and that does not require us to discount the elementary character of other colours. In particular, Jameson and D'Andrade (1997) give an explanation that seems to me plausible. They observe that Hering's elementary hues correspond to protrusions on the Munsell colour solid, which was used to generate colour samples in the WCS. The Munsell system (unlike the colour solids of Runge and the NCS) encodes experienced differences in colour in terms of distance in colour space. The protrusion of Hering's elementaries from the body of the Munsell colour solid, reflects the

[13] In fact, it is not entirely clear that BCTs for Hering's elementaries are always introduced before others. For some apparent exceptions, see Lindsey et al. (2020, 19); Davidoff et al. (1999).

fact that in their unsaturated forms they are recognized as more different to other colours than are, say, the colours in areas around the central body of the colour solid. But recognising a colour as different from another does not necessarily depend on whether they are different elementary colours. A saturated red and unsaturated red may lie at some distance in the Munsell colour solid, but an unsaturated green and unsaturated red may lie very close together. The experience of significant colour difference is not necessarily an indicator of a colour's elementary character, which is also to say that some elementary colours may be more visually obtrusive than others (such as Hering's elementaries), and others, especially those closer to the core of the Munsell colour solid may appear to us as rather more subtle (such as brown, and perhaps pink and grey).[14]

So, how many elementary colours would there be, using the basic color terms as a guide? If we accept that the BCTs in most languages with well-developed colour vocabularies are good indicators of which colours are elementary, that suggests that there are eleven elementary colours. In addition to Hering's elementaries, they are: brown, orange, purple, pink and grey. I have focused on orange, purple and brown in part because it seemed easier to persuade those I have spoken with, and myself, that these colours are elementary. Intuitions seem less clear on pink and grey. Is grey, as Hering says, simply a phenomenal mixture of black and white colour? Maybe, but some others do not to see it that way. The art historian Frances Guerin, in her book on grey in modern painting, is clear that for her it has a distinctive appearance of its own (Guerin 2018). Colour theorist Faber Birren observed, "[c]ertainly there are few, if any, persons who, upon looking at a gray area, will describe it as being a whitish black or a blackish white" (Birren 1969, 49). Is pink a whitish reddish colour? Personally, I have often felt that it is no more than this. But I can find few references to it in these terms, except when speaking of mixing paints, so I wonder if my experience is unusual. Boynton, for one, suggests there is "[p]erhaps … something slightly more unique [i.e., elementary] about it than can be fully described in terms of its … components" (Boynton 1975, 317).[15]

One might choose to take those languages that recognize twelve BCTs as a guide, dividing blues into two elementary colours, one corresponding to light blue and the other to dark blue. This raises the possibility that members of different language groups may experience different colours as elementary. This is suggested by a study that shows that speakers of Russian are better able to distinguish light and dark blues than English speakers (Winawer et al. 2007). Such a cultural variation in the number of elementary colours would imply a plasticity in how colour information is

[14] It could be objected that I should not be making use of orthodox colour spaces, such as the Munsell color solid, since some of the colours I have discussed—the stygian hues—do not fit into this solid. However, as Churchland (2005) describes, places for these colours can be found by extending existing colour spaces (i.e., by adding further protrusions to them), and this will allow the argument to go through without adjustment. In Chap. 6, I will say a little about colour spaces with more dimensions, but this too does not affect the argument made here (the addition of these further dimensions will not affect the distances between points in the existing dimensions).

[15] I am struck too by Li et al.'s description of pink as being different to red in colour rather than in lightness (Li et al. 2014, 215). I return to this in Chap. 4.

encoded in the brain, and require that culture, environmental conditioning, or some combination of the two, can play a role in shaping that coding.[16] I do not aim to settle on one of these positions here, though, as will become apparent in Chap. 5, I see no reason why the plasticity of the visual system could not allow cultural influences to partly shape our perception of elementary colours. In any case, linking elementary colours to BCTs is only a suggestion. It would not surprise me if it turned out to be a helpful one—since it is reasonable to think that the elementary colours will be particularly obtrusive and salient, and so likely to earn names in preference to other colours. But there is no necessary reason why that should be so. Ultimately, we should take experience as the final arbiter of whether a colour is elementary or not.

References

Androulaki A, Gômez-Pestaña N, Jover JL, Coventry K, Davies I (2006) Basic colour terms in Modern Greek: twelve terms including two blues. J Greek Linguist 7:3–47

Berlin B, Kay P (1969) Basic color terms: their universality and evolution. University of California Press, Berkeley and Los Angeles

Birren F (ed) (1969) A grammar of color: A basic treatise on the system of Alfred H. Munsell. Van Nostrand Reinhold, New York

Boring EG (1942) Sensation and perception in the history of experimental psychology. Appleton, New York

Boynton RM (1975) Color, Hue and Wavelength. In: Carterette EC, Friedman MP (eds) Handbook of perception, vol 5, 310–347. Academic Press, New York

Brentano F (1907) Untersuchungen zur Sinnespsychologie. Duncker & Humblot, Leipzig

Churchland PM (2005) Chimerical colors: some phenomenological predictions from cognitive neuroscience. Philos Psychol 18:527–560

Conway BR, Malik-Moraleda S, Gibson E (2023) Color appearance and the end of Hering's opponent-colors theory. Trends Cogn Sci 27:791–804

Davidoff J, Davies I, Roberson D (1999) Colour categories in a stone-age tribe. Nature 398:203–204

Feynman R, Leighton R, Sands M (2013) The Feynman lectures. Basic Books, New York

Fuld K, Werner JS, Wooten BR (1983) The possible elemental nature of brown. Vision Res 23:631–637

Guerin F (2018) The truth is always grey: a history of modernist painting. University of Minnesota Press, Minneapolis

Hardin CL (2013) Berlin and Kay theory. In: Luo R (ed) Encyclopedia of color science and technology, 1–4. Springer, New York. https://doi.org/10.1007/978-3-642-27851-8_62-2

Hardin CL (1993) Color for philosophers: unweaving the rainbow, expanded. Hackett, Indianapolis

Hering E (1964) Outlines of a theory of the light sense (trans: Hurvich LM, Jameson D). Harvard University Press, Cambridge, Mass

Jameson KA, D'Andrade RG (1997) It's not really red, green, yellow, blue: an inquiry into perceptual color space. In: Hardin CL, Maffi L (eds) Color categories in thought and language. Cambridge University Press, Cambridge, pp 295–319

Kay P, Berlin B, Maffi L, Merrifield WR, Cook R (2009) The world color survey. CSLI Publications, Stanford

[16] However, I would reject the idea that the ability to use BCTs is necessary for us to experience corresponding colours. Among other implausible implications, that would presumably reduce the colour experience of animals to nothing.

Li M, Liu F, Juusola M, Tang S (2014) Perceptual color map in macaque visual area V4. J Neurosci 34:202–217

Lindsey DT, Brown AM, Lange R (2020) Testing the cross-cultural generality of Hering's theory of color appearance. Cogn Sci 44. https://doi.org/10.1111/cogs.12907

Matthen M (2021) Unique hues and colour experience. In: Brown DH, Macpherson F (eds) The Routledge handbook of philosophy of colour. Routledge, Abingdon and New York, pp 159–174

Newall M (2022) A study in brown. Synthese 200. https://doi.org/10.1007/s11229-022-03956-9

Ocelák R (2015) The myth of unique hues. Topoi 34:513–522

Padgham CA, Saunders J (1975) The perception of light and color. Academic Press, New York

Paramei GV (2007) Russian "blues": Controversies of basicness. In: MacLaury RE, Paramei GV, Dedrick D (eds) Anthropology of color: interdisciplinary multilevel modeling. John Benjamins, Amsterdam, pp 75–106

Paramei GV, D'Orsi M, Menegaz G (2018) Diatopic variation in the referential meaning of the "Italian blues." In: MacDonald LW, Biggam CP, Paramei GV (eds) Progress in colour studies: cognition, language and beyond. John Benjamins, Amsterdam, pp 83–106

Quinn PC, Rosano JL, Wooten BR (1988) Evidence that brown is not an elemental color. Percept Psychophys 43:156–164

Sharp WA (2024) Russellian representationalism and the stygian hues. Erkenntnis 89:777–797

Watt HT (1908) Review of F. Brentano, Untersuchungen zur Sinnespsychologie. Mind 17:128–129

Webster MA, Kay P (2007) Individual and population differences in focal colors. In: MacLaury RE, Paramei GV, Dedrick D (eds) Anthropology of color: interdisciplinary multilevel modeling. John Benjamins, Amsterdam, pp 29–53

Westphal J (1982) Brown. Inq: Interdiscip J Philos 25:417–433

Winawer J, Witthoft N, Frank MC, Wu L, Wade AR, Boroditsky L (2007) Russian blues reveal effects of language on color discrimination. Proc Natl Acad Sci USA 104:7780–7785

Wright W (2019) The unique hues and the argument from phenomenal structure. Philos Stud 176:1513–1533

Yendrikhovskij SN (2001) A computational model of color categorization. Color Res Appl 26:235–238

Chapter 4
Reddish Greens and Yellowish Blues

Abstract While there are no surfaces, paints or lights that are reddish green or yellowish blue, many of us, I show here, can experience these colours. This chapter includes examples of stimuli that are reported to produce experiences of reddish green and yellowish. This is especially striking in the context of the philosophy and science of colour, because these colour experiences are forbidden by the dominant theories of colour: Hering's account of colour experience, and the opponent processing theory.

Keywords Colour · Novel colours · Impossible colours · Reddish green · Yellowish blue

Hering places two bars on the mixture of his elementary colours: red and green cannot appear mixed in our experience, nor can yellow and blue. So, there can be no reddish greens or yellowish blues. "No color is clearly reddish as well as greenish, no colour is both yellowish and bluish at the same time" (Hering 1964, 49). There are indeed no surfaces, paints or lights that are coloured reddish green or yellowish blue. Hering's prohibitions also forbid the mixture of any colours that would include a mixture that includes red and green, or yellow and blue components according to his phenomenology. So, for example, Hering also holds that orange and blue can never appear mixed, since for Hering, orange is a mixture of red and yellow). This rule similarly forbids various other combinations of colours, including purple-orange, orange-green, green-purple and purple-yellow. And, indeed, there are no surfaces, paints or lights that show these colours, nor any other colours that include the illicit combinations of red and green, and yellow and blue. The primary aim of this chapter is to present a range of evidence showing that while there are no surfaces or lights that are reddish green or yellowish blue, they can indeed be seen. I expect the other impossible pairings of colours on Hering's account can also be seen, and we will find as we go along that this is the case for a few such pairs.

The range of impossible colour pairings raises an issue that I will signal now, and address properly at the end of this chapter. Hering explains the apparent impossibility of reddish greens and yellowish blues by proposing that they cannot be encoded by opponent processing. For Hering, that also elegantly explains the impossibility of

M. Newall, *Impossible Colours: Colour Experience, Vision Science, and the Philosophy of Perception*, SpringerBriefs in Philosophy,
https://doi.org/10.1007/978-3-032-15137-7_4

the other forbidden pairings, since they all have either reddish green or yellowish blue components. But the expanded range of elementary colours I argued for in the previous chapter may seem to close off that way of explaining this apparent impossibility. For example, in the case of orange-blue, since I count orange as an elementary colour, it does not have a yellow component. So why is orange-blue impossible in the same way as yellow-blue (there being no orange-blue surfaces or lights)? This is a problem that I will address at the end of the chapter. Suffice it to say for now that I will make use of opponent processing to explain the apparent impossibility of all these colour pairs. This involves understanding that while opponent processing does not directly correspond to our experience of colour, it feeds into and constrains higher-level processing which more closely accords with that experience.

Historically speaking, Hering's idea that we never experience these colour mixtures is distinctive. An observation by the painter, Runge, in his 1806 letter published in Goethe (1810), might seem to express the same thought. There, Runge says that thinking of a reddish green is like thinking of "a southwesterly north wind" (Kuehni 2008, 63). That may appear to straightforwardly describe a fact about experience (or the possibility of experience), but reading Runge's broader writings on colour, it is clear that he is primarily thinking about the physical mixture of paints and our experience of those. His "elementaries" are the painters' primaries, red, yellow and blue, and he does not place a bar on the mixture of yellow and blue, but rather accepts that they mix to make green.[1] Hering transforms the claim into one that is unequivocally about experience.

A brief word is merited about reddish green's place in an earlier philosophical tradition. Runge influenced Wittgenstein's (1977) thought on colour, and through Wittgenstein, philosophy more generally. Drawing on Runge, Wittgenstein (1977, I–21, III–94) took the absence of reddish green from our experience to be a fact, and his analysis concerns the nature of what kind of fact that is. For me, it is apparent that it is not a fact at all, and so I take Wittgenstein's analysis of the question of what kind of fact it is as moot. In this, I follow Arthur C. Danto in his foreword to Hardin (1993, ix–xiii).

Hering proposed that information about the opponent colours is encoded in channels such that the coding of one colour prohibits the coding of its opponent colour. In particular, there is one channel that encodes red and green so that, when cells are active it indicates one colour, and when they are inactive, it indicates the other. Similarly, another channel encodes yellow and blue, so that when its cells are active one colour is indicated, and when they are inactive, the other is indicated (Hering 1964, 50). Reddish green and yellowish blue are thus "impossible" or "forbidden" in the sense that their encoding is prohibited by these mechanisms.[2] As I have mentioned,

[1] This is not to say that Runge was only interested in pigmentary mixture, but he took pigmentary mixture as indicating the scope and limits of the colours that we are capable of experiencing. (He was also interested in additive mixture using spinning disks, with which he experimented (Kuehni 2008, 92–97)).

[2] Hering puts it as follows: "[w]e should conclude from this that in the inner eye a physiological process whose psychological correlate would be simultaneously both red and green or yellow and blue is either not possible at all or is possible under only quite special unusual circumstances"

neural mechanisms answering to this description—encoding colour information as it leaves the eye and enters the brain—were discovered by Hurvich and Jameson (1957), cementing Hering's phenomenology in the scientific study of vision. Impossible colours, as I call them—reddish greens and yellowish blues—are therefore usually held not to exist. For most vision scientists and contemporary philosophers who write about colour, they have no part in our colour experience, and cannot do so, since they are prohibited by the opponent process theory.

However, this opinion is not universal. In an article in *Science*, Crane and Piantanida (1983) claimed to induce experiences of reddish green and yellowish blue. To do this, they used eye trackers to stabilize samples of red and green, and yellow and blue in the subjects' field of vision. Stabilization causes perception of the boundary between the colours to fade, and as this study describes it, the visual system compensates for this failure by filling-in the missing percept with an impossible colour—reddish green, or yellowish blue. Crane and Piantanida describe the experience produced in their subjects by the stabilized red and green stimulus:

> After a few seconds of viewing, the stabilized boundary between the red and green stripes disappears. Under these conditions, observers report that the field may have one or more of three distinctly different appearances: (i) the entire field appears to be a single unitary color composed of both red and green; (ii) the field appears to be composed entirely of a regular array of just resolvable red and green dots; or (iii) the field may appear as a series of islands of one color on a background of another color. (Crane and Piantanida 1983, 1079)

It is, of course, the appearance of the entire field as both red and green that is interesting in this context. Crane and Piantanida describe the red and green components of this impossible colour as of "significantly diminished saturation" compared to the saturated red and green of the stimulus. Nevertheless, most of their subjects reported that "regardless of where they attended in the field, the color was simultaneously both red and green". (Ibid.) Crane and Piantanida were similarly able to induce a yellowish blue percept in some of their subjects using a yellow and blue stimulus. A repeat of the experiment by a different group has replicated the results (Billock et al., 2001). Other studies have reported inducing similar effects with interactive displays. Brenner and Cornelissen (1991, 72) describe a stimulus that subjects could adjust until they saw "a green 'fog' [that] appeared to float in front of a red background." (I will say more below about whether such transparency effects should be considered experiences of reddish green. For now, it is enough to note that the experimenters took it to involve a reddish green percept).[3] Livitz et al. (2011) overlapped neon colour spreading effects to produce a colour that subjects reported as both reddish and greenish. Among philosophers, Hardin (1993, 124–126) was cautiously open to Crane and Piantanida's claims, and found the results of Billock et al., and Livitz

(Ibid.). On the basis of this, Hering does not wholly rule out the kind of account I develop in this chapter.

[3] I thank Eli Brenner for bringing this to my attention. Brenner and Cornelissen's result is hidden at the end of their paper. Brenner observes that the effect "is probably [produced by] the same mechanism as in your painting example [i.e., Newall 2021]". (Eli Brenner, personal communication, 10 February 2021).

et al., to rule decisively in favour of the existence of impossible colours (Hardin 2014, 380–382). Macpherson (2003, 2021) is another philosopher who views Crane and Piantanida's study favourably.

I think it likely that the reception of these studies has suffered from the fact that readers mostly lacked the apparatus to reproduce these colour effects for themselves. Seeing, as they say, is believing. But a variety of other more accessible methods of inducing impossible colours have been proposed recently. In an article in *Scientific American*, Billock and Tsou (2010) suggest that looking at the stimulus reproduced as Fig. 4.1 can produce experiences of yellowish blue (and a similar image using red and green can produce an experience of reddish green). To obtain the effect, the viewer must stare at the image, allowing their eyes to cross so that the crosses align.[4] Billock and Tsou (2010, 63) say that "[s]ome people get glimpses of forbidden [colours] … but the method is much less reliable than using equiluminant stabilized images." This stimulus has been widely circulated on the internet, producing some striking responses. A popular science YouTube Channel, The Action Lab, has a 2016 video encouraging viewers to record their experiences.[5] At the time of writing, the video has attracted over 20,000 comments. Dipping into these, it is apparent that Billock and Tsou's stimulus produces a variety of effects. Some commenters see green or grey. Some see yellow and blue alternating, "flip[ping] between the colors" (@sbeasley1120), or interpenetrating, "[t]hey were mixed together but it was … a heterogeneous mixture. It was like mixing a bunch of yellow and blue rocks". (@kylemartell-grimslid2326). Some describe an experience of seeing one colour through another, transparent, colour. One writes that "it looked like blue overlaid by yellow but didn't turn green" (@sbeasley1120), and another observes, "I can imagine achieving a similar colour effect in real life … [by taking a] blue base and misting the yellow on top with an ultra-fine spray" (@jameswright9810). Some of these viewers seem to count these experiences of seeing blue through yellow (or vice versa) as involving an experience of yellowish blue, and—as I have said—I will come back to the question of whether or not they could be correct about this. In any case, many viewers straightforwardly claim to experience yellowish blue, and some of these give careful and unequivocal descriptions of their experiences. One writes, "I saw the two colors blending together like a vertical watercolor wash mixing, then at a certain point it really did seem yellowish-blue. It certainly wasn't green, or yellow, or blue" (Lorrainejessop7332).

I have found two kinds of striking anecdotal observations of impossible colours in the natural environment. One kind is found in descriptions of plants that seem both greenish and reddish, or greenish and orange. For example, an online nursery describes the leaves of its Cordylines as "bronze green" and "green-maroon" (https://www.australianplantsonline.com.au/). Figure 4.2 shows, at top, a botanical print of a

[4] Billock tells me that the image (together with the red-green version) is due to Jeff Hovis. The attribution was omitted from Billock and Tsou (2010) due to a copy editor's error. (Billock, personal communication, 24 April 2020).

[5] 'Why There Are No Bluish-Yellow Crayons: The Forbidden Color Experiment', The Action Lab, YouTube, https://www.youtube.com/watch?v=cQCsDfEqr9o.

Fig. 4.1 Stereoscopic stimulus used to induce perception of yellowish blue. Adapted from Billock and Tsou (2010, 63)

Cordyline Fruticosa, which gives a sense of the stimulus the nursery copywriter has in mind. Let me add two other examples, notable for their inclusion in scientific journals. Below in Fig. 4.2 is an image of mangrove calyces, from a recently discovered hybrid species. The authors of the paper in which it appears report the calyces sometimes appear "green with a reddish blush", and describe other parts of the plant as "pink-green" (Duke and Kudo 2018, 282–283). Another group of researchers describe the mangrove calyces of the species they are studying as "greenish red" and "greenish orange" (Ragavan et al. 2014, 1035, 1036).

Another kind of observation of impossible colours in the natural environment, which I will spend more time over, occurs in descriptions of the sky at dawn or dusk. When the sky is clear, it is sometimes described as gradating directly from blue to yellow or orange at the horizon. That is, it is said to grading through a subtle yellowish blue or orange-blue—an impossible colour—rather than a neutral grey. Philosopher Paul Nedelisky, aware that we are not supposed to perceive such colours, gives a vivid description of the effect:

> One evening, about fifteen years ago, I was waiting at a traffic light and staring absently at the sunset. It was a long light, and after I had stared at the same patch of sky for about a minute, I found myself mentally describing its color: "Yellowish-blue..." ... It was pure, smooth, yellowish-blue. The light changed, and I had to drive away from the sight. I knew the experience would stick with me. In fact it did, and not merely in my memory. Over the years, I found I could often see the mysterious yellowish-blue at edges of sunsets. Again and again, I would stare at the color, as if by sheer intensity of attention I might pass through the veil of unintelligibility." (Nedelisky 2015)

The art historian, James Elkins, has the same effect, in detailed records of his observations of the clear sky at sunset (Fig. 4.3) (Elkins and Fiorentini 2020, 115–118).[6] In the image at left, Elkins' notations record pale blue grading directly into pale orange. One might wonder whether Elkins has left out a subtle grey band in between blue and orange. However, other images in his project carefully record grey bands of

[6] An earlier version appears in Elkins (2000).

Fig. 4.2 Top: Cordyline Fruticosa, stipple engraving with hand colouring. From Pierre Joseph Redouté, *Les Liliacées,* 1802–16. Bottom: Mangrove calyces. Photograph Norman C. Duke. From Duke and Kudo (2018), Fig. 4(l)

Fig. 4.3 Bands of colour observed in a clear sky around sunset. Left: Looking east, before sunset. Right: Looking east just after sunset. From Elkins and Fiorentini (2020), Fig. 9.1 (details)

colour (ibid.), so it appears that he meant to record the blue and orange as blending together directly. The image at right simply indicates a band of "yellow-blue".[7]

I think that landscape painters can also capture these effects in depicting the sky at dawn or dusk. One of the earlier painters of such skies is the seventeenth century painter Claude Lorrain. Claude often painted scenes looking into the sun just before sunset or after sunrise. One sees this, for instance, in his *A Mediterranean Port at Sunrise with the Embarkation of Saint Paula for Jerusalem*, c. 1657, which shows a sky clear, but for a few small clouds, smoothly grading from hazy blue at the top, to pale yellow at the horizon. J. M.W. Turner is another painter who depicts the same kind of sky. *The Blue Rigi, Sunrise*, 1842, is a good example. Other notable instances are Charles Gleyre's *Evening or Lost Illusions*, 1843, and Philipp Otto Runge's masterpiece, *The Small Morning*, 1808 (Fig. 4.4). Each shows much the same kind of sky, blue at the top, merging smoothly into yellow close to the horizon.

A YouTube video by the painter Shaun Ryan shows his technique of painting this kind of sky (Fig. 4.5). I have found that the video allows viewers to focus on the image in a way that can effectively prompt reports of yellowish blue colours, and I developed an experiment around it to see how widespread perceptions of yellowish blues are. The results were published in the psychology journal, *Perception* (Newall 2021) which I draw on below.

Ryan begins with a band of blue at the top, a band of yellow (mixed with a little red to give it an orange tint) at the bottom, and white in between. He blends them together until the blue and yellow pigments delicately mix with the white paint.

I asked a group of eleven subjects to watch the video, pausing it once the colours were fully blended (as in the top image in Fig. 4.6). I then asked them to fixate on a point in the image and indicate how far up they perceived the yellowish tint on Ryan's painting, and how far down they perceived a bluish tint. Of the eleven observers, seven recorded that these yellowish and bluish areas overlapped to some degree and indicated that they saw a colour that was simultaneously yellowish and bluish in that area, rather than, say, intermingling of islands or pixels of different colours. A

[7] I doubt that everyone with normal vision sees these colours in the sky as Nedelisky and Elkins do. Though Elkins' observations are his own, the format of his study echoes that of the astronomer Marcel Minnaert, who does not record seeing yellowish blues. (Minnaert 1993, Fig. 169) See Newall (2021, 132), for more on this variation.

Fig. 4.4 Philipp Otto Runge, *The Small Morning*, oil on canvas, 1808

Fig. 4.5 Shaun Ryan, screenshot from how to paint a sky—Acrylic painting lesson, 2013, YouTube, https://youtu.be/Zr9kMc25bPI

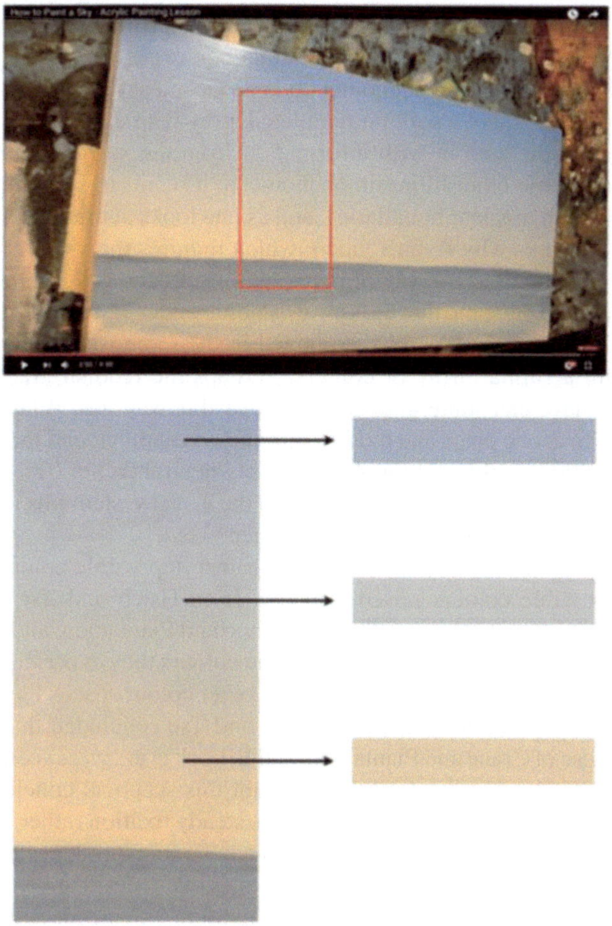

Fig. 4.6 Top: Still from Shaun Ryan, How to Paint a Sky—Acrylic Painting Lesson, taken at 3:50, with detail reproduced below outlined in red. Bottom: detail reproduced from still image above, with areas of colour from this detail isolated at right

number described their experiences of how the yellow and blue blended. One wrote that "in the middle they get intermingled", and another described them as "mixed". One subject described the colour as "yellow/blue pastel." The remainder said they experienced grey, green, or a transparency effect of yellow akin to those described by the commenters on The Action Lab video. Philosopher Fiona Macpherson (2021) includes an account of my study, concluding that she "leave[s] it to the reader to assess the merits of Newall's proposal, but for what it is worth, it appears phenomenologically plausible to me" (Macpherson 2021, 198).

There is, of course, no yellowish blue paint, and so the effect must be produced in some other way. As is clear from from the lower image in Fig. 4.6, the area that some see as yellowish blue is in fact grey—it has no chromaticity at all. So, the effect is an illusion of some kind, produced by the particular context in which the grey stimulus is presented. I suggest that the phenomenon is produced as follows. In the presence of a limited stimulus—here a soft, unmodulated, grey with unclear boundaries—the visual system fills the area in with information from the surrounding, yellow and blue, areas. Like some other filling-in phenomena, it seems to strengthen where the stimulus has smooth, unclear boundaries, and as one looks at the stimulus. The latter point perhaps explains why Ryan's video, which requires the viewer to spend time with it, is a relatively effective way of inducing this effect.

It is reasonable to presume that the yellowish blue effect that Nedelisky and Elkins observe in real skies at dusk can be explained in the same way, since they present the viewer with a similar array of colours. Perhaps the reddish green appearance that plants can have can sometimes be explained this way too. Where a plant, for example, is green at one place and reddish or orange at another, and there is a smooth transition between the colours through a relatively neutral colour (as is the case with the mangrove calyces in Fig. 4.2), it may be that the visual system fills in the neutrally coloured area with reddish green or orange-green.

I now turn to answering objections raised against impossible colours. The initial reports of impossible colours raised criticism from Hsieh and Tse (2006). They conducted an experiment similar to those of Crane and Piantanida, and Billock et al., finding that their subjects were able to match the colours they experienced to colours in CIE colour space. CIE colour space, like the other colour spaces I have discussed, does not include impossible colours, so Hsieh and Tse concluded that neither their subjects, nor those of Crane and Piantanida, and Billock et al., were seeing impossible colours. However, there is good reason to doubt this sceptical conclusion. Billock points out that Hsieh and Tse's experiment uses steady fixation rather than stabilization, and that, as Ditchburn (1972) observes, steady fixation produces weaker results. "Although I saw yellowish blues and bluish yellows many times under stabilization," Billock says, "I never saw them under steady fixation". (Vincent Billock, personal communication, 24 April 2020) Other studies that I have considered—Brenner and Cornelissen (1991), Livitz et al. (2011), as well as my own (Newall 2021)—do not use stabilization, but they can be defended from Hsieh and Tse's criticism in another way. In each of these studies, subjects were asked to distinguish the observed colours from the kinds of neutral colours (grey, and subtle shades of grey) that are most likely to be suggested as matches in CIE colour space. That strongly suggests that the subjects in these studies would in fact be able to distinguish their percepts from colours in CIE colour space and other orthodox colour spaces were they asked to do so.

I now turn to the topic of impossible colours and transparency. This arises for the following reason. We have already seen a number of transparency effects among descriptions of impossible colours. That is, rather than being what psychologists call a surface colour, one of the components is experienced as if seen through the other, as in Brenner and Cornelissen's description of "a green 'fog' [that] appeared to float in front of a red background." Could it be that other, perhaps all, examples of impossible colours are actually of this kind? For example, could the experiences of yellowish blue skies be of a transparent yellow haze in front of a blue sky? Now, when we happen to see a blue surface through a transparent yellow surface in life (say, the sky through a yellow pane of glass) we simply perceive the world as it is. On this line of thinking, we won't experience anything as yellowish blue, we just experience one coloured object (blue), behind another coloured surface (yellow), and experience both as having the colours they in fact have. Thus, such experiences of transparency do not involve experiences of impossible colours. And if all purported experiences of impossible colours turned out to involve similar perceptions of transparency, then none of the experiences I have discussed would truly be of impossible colours.

What should we make of this sceptical line of thought?[8] First, most of the experiences I have reported do not involve an experience of transparency, and if we can take at least some of these at face value, then some straightforward experiences of impossible colours do occur. In the case of the sky, one can also note that while it may give rise to experiences of transparent mists and hazes, it need not necessarily do so. The illusion of the sky as the 'vault of heaven', a distant hemispherical surface, is also a common experience. Still, I want to take time to examine the idea that the apparent impossibility of at least some colours can be dispelled by understanding them as transparency effects, for the links between impossible colours and transparency are interesting and illuminating.

In everyday experience, the transparency effects we are concerned with—such as seeing a blue object through a yellow surface, so that one remains visually aware of both colours—rarely occur. If one looks at the sky through a yellow pane of glass, one is unlikely to experience the view of the glass as involving a percept of blue and yellow simultaneously. Rather, it will tend to look a very dark colour, often a dark green. Looking through a blue glass at a yellow surface tends to produce the same effect. This is because the mixture involved when looking at a surface through transparent coloured glass is subtractive, like that involved when mixing paints. Looking through a red glass at a green surface, or through a green glass at a red surface, also rarely results in an experience of both colours. Instead, what often results is a view of a surface that appears a dark dull red, or a dark, dull green. So, most experiences of physical transparency that one might expect to give rise to an awareness of Hering's opponent colours do not actually do so. The most reliable and easy to study examples are seen in pictures, such as Fig. 4.7. (This is an example involving yellow and blue, but similar examples could be given using red and green, allowing a comparable analysis.) The image is the product of an exercise devised by the Bauhaus artist and teacher Josef Albers (1975, 24–26). Albers tasked his students to produce images of transparency where both colours retain their chromatic quality in the area of overlap. In this case, one not only has the impression of one sheet (e.g., the blue sheet) visible through other (e.g., the yellow sheet), one also simultaneously retains a visual awareness of both yellow and blue in the area of overlap.

It seems to me, at least, that the area of overlap does bear some similarity in its effect to the examples of yellowish blue I have already discussed. So, I think the effect is worth looking at more closely. The first point to make is that the area of

[8] Livitz et al. (2011, 10–12) consider this question and develop a response, though it is not one that can be generalized to all the cases I look at.

Fig. 4.7 Blue and yellow rectangles, showing an illusion of transparency in the area of overlap. Adapted from Albers (1963, ill. IX–1)

overlap, assessed as a surface colour, is in fact a neutral grey. As with Ryan's image, this can be seen most easily if the grey area is isolated from the colours that surround it. But when the grey area is seen in the context of the whole image, its appearance changes. It no longer appears as a grey surface, but instead as two coloured surfaces: a yellow surface through which is seen a blue surface, or vice versa. How does the appearance of this grey area transform in such a dramatic way? Here I draw on the analysis of the perception of transparency made by Fabio Metelli (1974) which has found wide application in vision science.

"[O]ne perceives transparency", Metelli says, "when one sees not only surfaces behind a transparent medium but also the transparent medium or object itself" (Metelli 1974, 91). He observes that physical transparency—a substance's ability to transmit light—is neither sufficient for producing the perception of transparency (as one may appreciate when one walks into a glass door), nor is it necessary for it (as Fig. 4.7 illustrates).[9] So, determining the conditions under which transparency is perceived is not a trivial matter, and, indeed, poses complex challenges for vision science. Metelli outlines a key element of a solution to this problem:

> With the perception of transparency the stimulus color splits into two different colors, which are called the scission colors. One of the scission colors goes to the transparent layer and the other to the surface of the figure below. ... there is a simple relation between the stimulus

[9] Metelli himself uses similar abstract configurations of shapes to explore the phenomenon of perceptual transparency.

> and the scission colors: when a pair of scission colors are mixed, they re-create the stimulus color (Metelli 1974, 93).[10]

Two terms require definition to make sense of this "law of scission". The first of these is *stimulus colour*, and second, is *colour mixture.* I find a recent approach to understanding stimulus colour and mixture suggested by computational neuroscientist Tony Vladusich to be particularly helpful here. (Vladusich 2013) This approach is best explained by first describing how the transparency effect seen in Fig. 4.7 is produced. I will come to revise this shortly, removing mention of colour sensations, but these will help in an initial explanation.

1. The visual system causes the edges of the grey area to appear yellowish (at left) and bluish (at right). This is the product of a simultaneous contrast effect which acts on the edges of the grey area. On its left side it is surrounded by blue, which causes the grey edges there to appear yellowish, and on the right side it is surrounded by yellow, which causes the grey edges at right to take on a bluish colour.
2. Next, a filling-in process occurs. The interior of the grey shape is filled-in, using the colour from the edges to fill-in the interior as a yellowish-bluish colour. This functions as the stimulus colour that will undergo scission according to Metelli's account. Presumably, this filling-in is similar to the filling-in processes already described by those who have studied impossible colours.
3. Then, a process of scission gives rise to the transparency effect, by "splitting" the yellowish-bluish colour so that its yellowish component is attributed to the transparent layer, and the bluish component is attributed to the underlying layer.

Attending carefully to Fig. 4.7, one can become conscious of these separate effects. Looking at the edges of the grey shape, one can see how the simultaneous contrast creates a coloured "glow" along the edges, that strengthens as one looks at them. Turning one's attention to the grey interior of the shape, one can see the filling-in effect establishing itself—the yellow on the left edges, and the blue on the right edges spreading into the interior, where they might appear to mingle. Or one might see the grey shape as both yellowish and bluish all over. As I look at the shape, I find that these effects come and go, the grey sometimes reappearing, and then the filling-in effect re-establishing itself. Once filling-in has achieved some level of completion, a transparency effect seems to be quickly established: an experience of a blue surface as if through a transparent yellow film.

Returning to Metelli's law of scission, it will be apparent that with the scission we experience in response to Fig. 4.7, the stimulus colour should be identified with the subjective colour that the stimulus presents (rather than say, its reflectance profile), and that "mixture" corresponds to phenomenal mixture. It is tempting to again slip into subjective terms here, as I have in my three-step description of the process, and say that the stimulus colour is yellowish-blue, and that this colour is split into its

[10] There are further conditions for transparency perception, which Metelli calls "figural". These stipulate how coloured shapes need to be configured to give rise to a transparency effect (Metelli 1974, 92–93).

yellow and blue components in transparency perception. I should now say that this kind of description is inadequate, for colours, *qua* subjective qualities, cannot play a role in the explanation of the process, aside from figuring in its output. I say this for two reasons. First, it is not clear what causal properties sensations of colour can have. Second, the three-step process I have described can occur unconsciously, that is, without any corresponding colour sensations experienced by the subject, aside from the final transparency effect. It is here that Vladusich's analysis of transparency perception is helpful, in that he allows a formulation of the concepts of stimulus colour and mixture in terms of computational states and processes that encode and compute information about colour.[11] (Vladusich 2013, 425) Accordingly, I restate my account as follows:

1. The visual system processes the edges of the grey area as yellowish (at left) and bluish (at right). This is the product of a simultaneous contrast effect which acts on the edges of the grey area. On its left side it is surrounded by blue, which causes the adjacent grey edges to be computed and encoded as yellowish, and on the right side it is surrounded by yellow, which causes the grey edges at right to be coded as bluish.
2. Filling-in occurs. This involves drawing on the information about the edges to compute and encode the interior as a yellowish-bluish colour.
3. A process of scission gives rise to the transparency effect, through a computation that produces and encoding that attributes the yellow and blue components of the yellowish-bluish colour to the transparent layer and underlying layer, respectively.

We are now able to return to the question of whether seeing blue through yellow involves seeing yellowish blue. On the account I have given, the visual system encodes yellowish blue as the outcome of step (2). So, regardless of whether the state is conscious or not, like impossible colours it is "impossible" in the sense that it cannot be encoded by opponent processing. Could this encoding occasion a corresponding experience of yellowish blue? The account I have just given leaves this open. It need not produce a perception of yellowish blue, but equally, it allows the possibility. But I think there are good reasons to expect that the experience of yellowish blue can emerge into consciousness under these circumstances. Consider, first, this famous passage from Ruskin.

> [T]he whole technical power of painting depends on our recovery of what may be called the *innocence of the eye*; that is to say, of a sort of childish perception of these flat stains of colour, merely as such, without consciousness of what they signify, as a blind man would see them if suddenly gifted with sight. (Ruskin 1857, 6, n. 1)

---

[11] Vladusich's theory is an advance on models of transparency perception that see it as the inverse of physical colour mixture, which is to say, additive fusion (episcotister or "spinning disk") mixture and subtractive (filter) mixture (e.g., Metelli 1974; Beck 1978; Beck et al. 1984; D'Zmura et al. 1997). "Unlike vision models based exclusively on the ecological analysis of physical constraints imposed upon the visual system by the outside world, such as the episcotister and filter models of transparency", Vladusich "suggests that the properties of perception need to be assessed in terms of the computational properties of the brain, not just the physical properties of the outside world" (Vladusich 2013, 425).

Ruskin is mistaken that this perception is primary (see, e.g., Gombrich 1960, 250–256), but, as art practice shows, he is correct that this kind of perception can be fostered. We can attend to the world as if it is a picture, as "flat stains of colour", and this is something that painters can become especially skilled at doing. If one can see one's visual environment as a picture, one can see what shapes and colours one needs to use in order to paint it—hence Ruskin's concern with this form of vision.[12] That suggests it may be possible to bring this ability to Fig. 4.7, and see it not as one surface visible through another, but as "flat stains of colour". How will the area that depicts the overlap look when seen this way? It might simply look grey, especially if one isolates it from the surrounding colours—but that is not the only possibility. As most painters know, when one attends to the visual environment in this way, one still remains subject to many optical illusions, including simultaneous contrast and filling-in. (This is important in painting, since one falls into errors of exaggeration if one tries to paint such effects in—instead, the painter usually needs to focus on accurately rendering the colours and forms that *produce* such illusions.) So, it is reasonable to think that simultaneous contrast and filling-in can also make themselves felt when one takes the "painter's perspective", and this will allow the perception of a yellowish blue. That, also, is what my own experiences of Fig. 4.7, described above, suggest: that careful attention can allow the effects of simultaneous contrast and filling-in—and the accompanying yellowish blue colour—to emerge into consciousness independently of a transparency effect.

Metelli's discussion of transparency perception also suggests that this kind of perception of the stimulus colour may be possible. Metelli notes that transparency perception can alternate with a veridical perception of the stimulus colour: "if the region of superimposition [is isolated] (even if it is just by the attitude of the observer), then only the [stimulus] color is perceived" (Metelli 1974, 91). This can be accomplished by physically separating the colour from its context (e.g., by covering up the surrounding colours), or just by changing "the attitude of the observer"—attending closely to the colour of the area of overlap at the expense of the surrounding colours. Again, we may ask if this perception will show us the shape as grey or as yellowish blue. If the surrounding coloured areas are covered up, it will appear grey. But if they are not, it is reasonable to think that simultaneous contrast and filling-in may still work, in which case a perception of yellowish blue will be produced. That is all to say that Metelli's account converges with that I have developed above.

A similar analysis to that I have given above can be furnished for experiences of transparency involving red and green. Figure 4.8 shows red and green rectangles appearing to overlap, and it will be apparent how my analysis will run here. One Quora user, posting an image similar to this, asks "Isn't this a reddish green?" (Fid Rewe, https://www.quora.com/Why-cant-we-see-a-reddish-green-color/answer/Fid-Rewe) I am not sure whether this is the best example of a red–green transparency effect that could be achieved, but it persuades some that

[12] This is also the essential insight of Betty Edwards' bestselling book, *Drawing on the Right Side of the Brain* (Edwards 1979).

Fig. 4.8 Configuration reported to induce "reddish green" perception. Adapted from Fid Rewe, comment on Quora, 2020. https://www.quora.com/Why-cant-we-see-a-reddish-green-color/answer/Fid-Rewe

they are perceiving a reddish green—and does so without the benefit of the kind of arguments I have presented above.

So, there are reasons to think that transparency effects can in fact allow the perception of impossible colours. One may also wonder whether the process I have outlined can be stopped at step (2) in a more decisive way, by designing a configuration of colour that prompts steps (1) and (2), but not (3). I think it may be. In Fig. 4.9, a grey rectangle is placed in the centre of a colour gradient that is red at one end and green at the other. Here, the red–green gradient produces a simultaneous contrast effect, so that the edges of the grey rectangle are processed as greenish, at left, and reddish, at right. The inner part of the grey rectangle is then subject to a filling-in effect, so that the left side appears greenish, and the right side appears reddish. Staring at it for a while, the green and red can appear to mingle in the middle. Figure 4.9 is adapted from an image by Minjal Dharia, who calls the resultant colour "a greenish red" or "reddish green"[13] (Dharia 2010). My sense is that if one can see this colour, it is a subtle effect, but the fact that some see it is another piece of evidence that the process can be halted before a transparency effect is produced, resulting in a straightforwardly reddish green percept.[14]

Finally, let me turn to another objection that can be raised against impossibly coloured percepts. Philosophers Martine Nida-Rümelin and Juan Suarez suggest

[13] Dharia's image, which she records was made in a colour theory class taught by the painter Michelle Henebrook, at the Pratt Institute New York, is an ingenious extension of Johannes Itten's colour demonstrations in which he uses simultaneous contrast to make grey squares take on different colours by placing them on coloured grounds (Dharia 2010; Itten 1973, 86–87).

[14] This stimulus does not create a transparency effect because we cannot see the grey rectangle as the overlap of two surfaces (that is to say, it does not satisfy Metelli's "figural" conditions for transparency perception).

Fig. 4.9 Stimulus reported to induce a "greenish red" or "reddish green" perception. Adapted from Minjal Dharia (2010), Contrast of complements (digital study), https://colorsynesthesia.wordpress.com/2010/11/09/juxtaposition-session-10/

that reported instances of reddish green differ in an important way from ordinary phenomenal mixtures of colours for this reason: while they appear simultaneously reddish and greenish, they do not really appear reddish green. As they put it, "red and green [can] be simultaneously present on a surface in a given experience without phenomenally mixing into a new hue" (Nida-Rümelin and Suarez 2009, 363). This is how they describe the experience:

> [It is] like … looking through a green glass at a red surface but the two surfaces do *not* appear to be at different places … there is only one apparent surface present in the experience. But still green is on the surface and red is on the surface too. However red and green are there (apparently) on the surface without mixing into one new colour. (Ibid.)

The experience, a footnote in their paper makes clear, is Nida-Rümelin's. She "discovered that experiences of [this kind] … occur (or can occur) when looking through goggles with a red and a green glass on a white surface" (Rümelin and Suarez 2009, 364, n. 28). The paper adds that Nida-Rümelin's co-author, Suarez, was "quite skeptical about this claim" (ibid.), and presumably for that reason they go no further than describing it as a "speculation" (362). Nevertheless, I think there are good reasons to take Nida-Rümelin's account of the experience seriously. First, the binocular combination of red and green images is similar to the strategy Billock and Tsou's stereoscopic image (Fig. 4.1) makes use of, and so it would not be surprising to find it yields experiences similar to the impossible colours user that we have seen users of that image report. Second, Kilian-Roy Lachner, in his experiments inducing artificial tetrachromacy, has systematically recorded his experiences of wearing glasses that use similarly (although not identically) coloured filters to those of Nida-Rümelin

(Lachner, n.d.).[15] He has worn these glasses for a long period of time, and describes in detail the impossible colours he observes.

What should we make of Nida-Rümelin's description of her red-green experience—and perhaps, by extension, those of others who use binocular combination to produce unusual colour experiences? It seems to me that Nida-Rümelin finds missing is a quality similar to Matthen's unity. Recall that for Matthen, "we experience each colour, whether simple or compound as *one* phenomenal whole, not as many phenomenal qualities bound together" (Matthen 2021, 166). Nida-Rümelin's red-green experience lacks that phenomenal wholeness or unity. As I said in Chap. 3, on my account, unity can be left out of a phenomenology of colour that allows non-Hering elementaries. Accordingly, I am inclined to say that the red-green that Nida-Rümelin describes, so far as it involves simultaneous experience of two elementary colours in the same place, in fact has all the characteristics required to count as an experience of a single colour, reddish green, and that this is as much a single colour as other phenomenal mixtures of elementary colours. I suspect that Nida-Rümelin's sense that something is lacking from her experience comes from comparing it with colours such as orange or purple, which on the orthodox account are counted as combinations of other colours. From the orthodox point of view red and yellow really do seem to undergo some transformation when (supposedly) phenomenally combined into orange. But on my approach, this is not so, since orange is not in fact a reddish yellow, but an elementary colour in its own right. So, I think that Nida-Rümelin may expect too much from the phenomenal combination of red and green. It would be better to compare her experience to a genuine phenomenal mixture on my account, such as greenish blue. As she describes her red-green experience, "green is on the surface and red is on the surface too." Much the same, I would say, is true of greenish blue—there, blue is on the surface, and so, at the same place is green. That, and no more than that is what the experience of greenish blue is, and I hold that much the same is true of seeing reddish green. It is simply an experience of the co-location of red and green on a single surface.

To sum up the results of this chapter, there are many plausible reports of impossible colours—reddish greens and yellowish blues—in both scientific and more informal contexts. Even in cases involving transparency effects, where one of Hering's opponent colours is reported as being seen through another, there are still strong reasons to think that experiences of impossible colours can occur. Thus, Hering's bar on the phenomenal mixture of red and green, and yellow and blue, should be rejected.

* * *

At the start of this chapter, I promised to explain why there are various other "impossible" pairings of colours, beyond reddish green and yellowish blue. For instance, like those colours, we never see orange-blue surfaces or lights, though, like yellowish blue, we can sometimes see it in the sky at dusk. Hering will explain orange-blue's "impossibility" by pointing out that orange-blue contains yellow and blue components (since for Hering orange in reddish yellow) and so cannot be encoded in opponent processing. However, for me, orange, as we experience it, need not

[15] Lachner's glasses are similar to those described in Gundlach et al. (2018).

contain any yellow component—so how can I explain why, like reddish green and yellowish blue, it does not seem to be a feature of surface and illumination colours? Other colours will raise the same problem for me, such as orange-green (seen in the mangrove calyces) as well as green-purple and purple-yellow. Why are these all as impossible as reddish green and yellowish blue—never colouring in surfaces, paints and illumination—despite not containing those colours in my phenomenology?

To solve the problem, we need to make a distinction between two stages of visual processing: the encoding of colour information at the stage of opponent processing, and that which occurs within the visual cortex, Opponent processing channels feed into structures in the visual cortex which encode colour information in a different way, more fully corresponding to our experience of colour by encoding the richer array of elementary colours that I described in Chap. 3, as well as "impossible" colours. However, by feeding into these structures, opponent processing still shapes our experience in important and distinctive ways.

How? An active red–green channel codes either red or green, but never both. This feeds into the visual cortex, and there that content is transferred into a new form. Activation of a "red" signal in the opponent channel, in the absence of input from the yellow–blue opponent channel, will activate a representation for "elementary red" in the cortex. If there is also input indicating "yellow" in the yellow–blue channel, that will activate a representation for "elementary orange" in the visual cortex. In this way, opponent coding as described by Hering is translated into a richer form of coding in the visual cortex, which comes closer to corresponding to our visual experience of colour. This two-stage process in which opponent coding is translated into another form of coding in the visual cortex can thus encode the larger range of elementary colours as I have described, while at the same time excluding all the colours Hering identifies as impossible: not only reddish greens, and yellowish blues, but orange-blues, and so on. Because those colours cannot be coded in opponent processing, they cannot be translated from the opponent processing channels into cortical representations. For this reason, impossible colours cannot be associated with a direct translation of information from the opponent processes. Rather, their encoding must be the product of different kinds of activity within the visual cortex—in particular, the neural processes corresponding to filling-in—which supplement the input from opponent processes. These processes in the visual cortex, allowing the encoding of the broader range of elementary colours I have described, as well impossible colours, are the topic of the next chapter.

References

Albers J (1975) Interaction of color. Revised edn. Yale University Press, New Haven, Conn

Albers J (1963) Interaction of color. Complete edn. Including portfolio of illustrations. Yale University Press, New Haven, Conn

Beck J (1978) Additive and subtractive mixture in color transparency. Percept Psychophys 23:265–267

Beck J, Prazdny K, Ivry R (1984) The perception of transparency with achromatic colors. Percept Psychophys 35:407–422

Billock VA, Gleason GA, Tsou BH (2001) Perception of forbidden colors in retinally stabilized equiluminant images: an indication of softwired cortical color opponency? J Opt Soc Am A 18:2398–2403

Billock VA, Tsou BH (2010) Seeing forbidden colors. Sci Am 302:58–63

Brenner E, Cornelissen FW (1991) Spatial interactions in color vision depend on distances between boundaries. Naturwissenschaften 78:70–73

Crane HD, Piantanida TP (1983) On seeing reddish green and yellowish blue. Science 221:1078–1080

Dharia M (2010) Color synesthesia (blog). https://colorsynesthesia.wordpress.com. Accessed 21 Aug 2023

Ditchburn RW (1973) Eye-movements and visual perception. Clarendon, Oxford

Duke NC, Kudo H (2018) Bruguiera × dungarra, a new hybrid between mangrove species B. exaristata and B. gymnorhiza (Rhizophoraceae) recently discovered in north-east Australia. Blumea—Biodivers Evol Biogeogr Plants 63:279–285

D'Zmura M, Colantoni P, Knoblauch K, Laget B (1997) Color transparency. Perception 26:471–492

Edwards B (1979) Drawing on the right side of the brain: a course on enhancing creativity and artistic confidence. J. P. Tarcher, Los Angeles

Elkins J (2000) How to use your eyes. Routledge, New York

Elkins J, Fiorentini E (2020) Visual worlds: looking, images, visual disciplines. Oxford University Press, New York

Gombrich EH (1960) Art and illusion: a study in the psychology of pictorial representation. Phaidon Press, London

Gundlach BS, Frising M, Shahsafi A, Vershbow G, Wan C, Salman J, Rokers B, Lessard L, Kats MA (2018) Design considerations for the enhancement of human color vision by breaking binocular redundancy. Sci Rep 8:1–9

Hardin CL (2014) More color science for philosophers. In: Stokes D, Matthen M, Biggs S (eds) Perception and its modalities. Oxford University Press, New York, pp 379–390

Hardin CL (1993) Color for philosophers: unweaving the rainbow, expanded. Hackett, Indianapolis

Hering E (1964) Outlines of a theory of the light sense (trans: Hurvich LM, Jameson D). Harvard University Press, Cambridge, Mass

Hsieh P-J, Tse PU (2006) Illusory color mixing upon perceptual fading and filling-in does not result in "forbidden colors." Vision Res 46(14):2251–2258

Hurvich LM, Jameson D (1957) An opponent-process theory of color vision. Psychol Rev 64:384–404

Itten J (1973) The art of color: The subjective experience and objective rationale of color (trans: van Haagen E). Van Nostrand Reinhold, New York

Kuehni RG (2008) Philipp Otto Runge's Color sphere: a translation, with related materials and an essay. Ebook. https://www.web3.lu/download/RungeFarben-Kugel.pdf. Accessed 16 Aug 2025

Lachner K-R. n.d. True-red non-retinal (chromatically-less-redundant) moderately functional tetrachromacy. https://www.color-in-color.info/tetrachromacy_1/non-retinal-tetrachromacy. Accessed 5 Apr 2025

Livitz G, Yazdanbakhsh A, Eskew RT, Mingolla E (2011) Perceiving opponent hues in color induction displays. Seeing Perceiving 24:1–17

Macpherson F (2021) Novel colour experiences and their implications. In: Brown DH, Macpherson F (eds) The Routledge handbook of philosophy of colour. Routledge, Abingdon, pp 175–209

Macpherson F (2003) Novel colours and the content of experience. Pac Philos Q 84:43–66

Matthen M (2021) Unique hues and colour experience. In: Brown DH, Macpherson F (eds) The Routledge handbook of philosophy of colour. Routledge, Abingdon and New York, pp 159–174

Metelli F (1974) The perception of transparency. Sci Am 230:90–98

Minnaert MGJ (1993) Light and color in the outdoors (trans: Seymour L). Springer-Verlag, New York

Nedelisky P (2015) Impossible wonder: a reply to James McWilliams. The Hedgehog review: critical reflections on contemporary culture. https://hedgehogreview.com/web-features/thr/posts/impossible-wonder-a-reply-to-mcwilliams. Accessed 6 Feb 2023

Newall M (2021) Painting with impossible colours: some thoughts and observations on yellowish blue. Perception 50:129–139

Nida-Rümelin M, Suarez J (2009) Reddish green: a challenge for modal claims about phenomenal structure. Philos Phenomenol Res 78:346–391

Ragavan P, Saxena M, Saxena A, Mohan PM, Sachithanandam V, Coomar T (2014) Floral composition and taxonomy of mangroves of Andaman and Nicobar Islands. Indian J Geo-Mar Sci 43:1031–1044

Ruskin J (1857) The elements of drawing; in three letters to beginners. Smith, Elder & Co, London

Von Goethe JW (1810) Zur Farbenlehre. J. G. Cotta, Tübingen

Vladusich T (2013) A reinterpretation of transparency perception in terms of gamut relativity. J Opt Soc Am A 30:418–425

Wittgenstein L (1977) Remarks on colour, Anscombe GEM (eds), (trans: McAlister LL, Schättle M). Blackwell, Oxford

Chapter 5
Beyond Opponent Processing

Abstract The previous chapters have shown more elementary colours are experienced than the orthodox theory recognizes, and that there are credible reports of colour combinations denied by the orthodox view of colour experience—reddish greens and yellowish blues. These results give a much richer phenomenology of colour than that of the orthodox view, and more accurately reflect the character of colour experience. But this account of colour experience meets a challenge in the opponent process theory, which prohibits these colour experiences elementary colours beyond those proposed by Hering, along with reddish greens and yellowish blues. This chapter responds by showing that the phenomenology of colour I outline finds support in recent scientific research on the neurophysiology of vision that suggests the possibility of a kind of coding of colour information that differs from opponent processing.

Keywords Colour experience · Theories of colour processing · Opponent processing · Unique colours · Reddish green

We have now reached two broad conclusions. First, there are elementary colours unaccounted for by Ewald Hering (1964) but which nonetheless are part of the experience of those of us with normal vision. Second, the restrictions Hering there places on phenomenal colour mixture do not in fact hold—that is to say, it is possible to experience reddish greens and yellowish blues. These two conclusions are incompatible with the opponent process theory, which only has the resources to encode Hering's six elementary colours, and is unable to encode reddish greens and yellowish blues. As we have seen, this is the reason why reddish greens and yellowish blues are known as impossible colours, and why the non-Hering elementaries can equally be called impossible.

Let me spell out the logical implications of this incompatibility. If the opponent process theory, understood as an account of how colour experience is determined, is correct, then none of my conclusions—none of the claims for the existence of non-Hering elementary colours and impossible colours—can be right. It also follows that if my two conclusions are sound, our experience of colours cannot be determined

M. Newall, *Impossible Colours: Colour Experience, Vision Science, and the Philosophy of Perception*, SpringerBriefs in Philosophy,
https://doi.org/10.1007/978-3-032-15137-7_5

by opponent coding. Thus, much depends on the possibility of finding an account of how colour experience is determined that is not bound by the constraints of opponent processing. There is, it has been suggested, scope to see a new paradigm in development (Conway et al. 2023), and this chapter introduces one approach to this, and outlines its relevance in the context of my phenomenological findings.[1] In particular, I will explain how this approach could plausibly allow the encoding of elementary colours beyond Hering's six, and of reddish greens and yellowish blues.

It is already thought by many vision scientists that the coding that occurs in opponent process pathways does not in fact correspond to our experience of elementary colours. Recall here the views of Mollon and Jordan, and Valberg:

> [M]ost colour scientists are agreed that the chromatically opponent cells of the early visual system ... do not correspond colorimetrically to red-green and yellow-blue processes. (Mollon and Jordan 1997, 382)
>
> [Opponent processing cells at this early stage are] not the correlate for seeing a particular hue quality, say unique [i.e., elementary] red. This latter correlate, if it exists as a separable entity, must be associated with yet unidentified, higher-level neural activities (Valberg 2001,1645).[2]

What, then, does determine our experiences of elementary colours? There are two general approaches. One holds that the visual environment conditions the sensitivities of the visual system, and this causes it to become especially responsive to the hues we experience as elementary. As Chauhan et al. put it:

> Perhaps this is the reason why the neural reality of perceptual red-green and blue-yellow hue-opponent mechanisms has proven to be so elusive—it is not a fundamental mechanism hard-wired into the neural circuitry, but a statistical peak in the tuning of neural populations, many of which multiplex both colour and luminance information [i.e., encode this information in a single signal]. (Chauhan et al. 2023, 12)

On this view, which Mollon (2006) dubs the "ecological" account, the elementary hues should primarily be understood as a feature of our visual ecology, rather than depending on the form in which colour information is encoded in the brain. But he acknowledges that it has shortcomings. While elementary blue and yellow may be important features of our visual environment (since they tend to track sunlight and shade), it is less clear why red and green would figure as elementary colours on this approach (Mollon 2006, 306–307). Noting this, Matthen, concludes that the "ecological account seems so far to rest on a few suggestive 'coincidences'" (Matthen 2021, 170). The situation would hardly be helped by the additional elementary colours that I have proposed—what universally salient features of the environment could orange, purple and brown track? The ecological approach also does not explain our

[1] My approach, I think, is consistent with the neuroscientific theory outlined by Conway et al. (2023), which I discuss in a footnote below.

[2] For a summary of the research that forms the basis for these conclusions, see Matthen (2021, 169), and for a clear argument for the position based on experimental findings, see Wuerger et al. (2005). Note that I do not intend to make any claim about neural correlates of consciousness (see my remarks below on this), and I doubt that Valberg means his use of "correlates" to be understood in that sense either. Rather, I understand a correlate to be a neural state bearing information that correlates with our colour experience, not one that is also sufficient to produce a colour experience.

capacity to experience impossible colours, since these still require a non-opponent method of encoding if we are to experience them, and the ecological approach says nothing about what form that may take.

The second approach looks for locations in the brain that show evidence of physically bearing content about colour experience beyond the opponent processing of the earlier visual system. Mollon (2006) calls this approach "constitutionalist", since it understands the elementary colours to be determined by distinctive features of processing or coding of colour information—that is to say, by features of the constitution of the subject's visual system. The most successful of those who take this approach are Stoughton and Conway (2008) and Forder et al. (2017).[3] However, critics have argued that their data are more appropriately interpreted as evidence of opponent processing playing the key role determining which colours are elementary (Mollon 2009, Chauhan et al. 2023).[4] Moreover, while Stoughton and Conway find a preponderance of cells in areas of V4 responsive to red, green and blue, their results for yellow are markedly less strong (Mollon 2006; Matthen 2021, 169). So, at present, "the neural basis of … the unique hues … remains an open problem" (Chauhan et al. 2023, 1).

I wonder if those researchers who adopt a constitutionalist approach are hampered by Hering's phenomenology. That is, in searching for bearers of the content of colour experience, they search for areas of the visual system that turn on in response, and only in response, to red, green, yellow and blue. If the account I have given of elementary colours is correct, they would be better seeking a broader range of bearers corresponding to a broader range of elementary colours. Indeed, I think that the focus on Hering's elementaries has caused researchers to largely overlook some recent relevant research—research that forms the basis of the account which I outline now. I will return to the debate between the ecological and constitutionalist views in Chap. 7. But it is worth signalling my position here. I see scope to reconcile the insights that motivate these two contending views. In particular, I understand elementary colours as determined directly by the constitution of the visual system, and indirectly by its visual ecology.

* * *

Information about colour is initially encoded by three kinds of photoreceptors in the retina, the S, M and L cones. Within the retina, this is translated into opponent processing pathways, coding black and white, red and green, and yellow and blue. The signals conveying this information exit the retina, travel through the optic nerve and Lateral Geniculate Nucleus, and enter the visual cortex. From V1, signals run to V2 and then to V4, which has often been regarded as the primary centre of colour processing (Zeki 1980). From there, colour signalling runs onward through the ventral stream of processing where it is integrated with other information, allowing functions such as object recognition. It is also clear that much information is sent "downwards"

---

[3] Kuehni (2012, 284) gives a helpful survey of some of this work.

[4] Mollon (2009) makes this criticism of Stoughton and Conway (2008). Chauhan et al. (2023) use it to counter Forder et al. (2017).

in the other direction, and this is also thought to play an important role in resolving more complex colour phenomena, such as transparency effects (Tsotsos et al. 2008). What then happens in these stages of colour processing in V1, V2 and V4, and how could they encode colour information in a form beyond that of opponent processing?[5]

In a number of studies, fMRI scans and two-photon imaging of macaques have shed some light on this question, providing detailed data about how cells in V1, V2 and V4 respond to particular colour stimuli.[6] These responsive areas are arranged in irregularly shaped layers of neurons which I call "patches", following Li et al. (2014). Patches have been found in V1 (Chatterjee et al. 2021), V2 (Xiao et al. 2003), and V4 (Li et al. 2014), and in all these regions by Liu et al. (2020).[7] Patches are arranged into columnar structures and each of the patches in such a structure is sensitive to colour in the same small part of the visual field. Strikingly, neighbouring patches are sensitive to hues that lie next to each other in the colour wheel. Xiao et al. (2003), were the first to image these structures, and show how they responded to the chromatic colours:

> The regions of the peak responses formed 70–320-nm-wide bands [i.e., patches] with various shapes, ranging from linear to nearly circular. Along these bands, nearby regions represent colours of similar hues, and a full cycle of colours is represented by an approximately 1.3-mm-long segment along the bands. (Xiao et al. 2003, 536)

Figure 5.1 shows the sensitivities of the patches running from blue, through green, orange, yellow, red and purple, and back to blue. The columnar structure bends around like a worm, its cross-section growing, distorting and shifting as it twists about.

These discoveries, striking as they are, were not unexpected. These structures are examples of what neuroscientists call a "hypercolumn", a feature made up of layered structures of neurons—"patches" in the terminology I use, but also called "columns" in this context. The hypercolumn functions as "a small machine that looks after all values of a given variable." (Hubel and Wiesel 1974) Each layer is sensitive to a different quality of a circumscribed kind, and together they are

---

[5] The answer that I develop here is consistent with an approach developed by Conway et al. (2023) They propose that the visual cortex codes colour categories (which I see as related to elementary colours) based on their utility in the context of the subject's physical and cultural environment. They posit that this coding may develop through the visual cortex's operation as a neural network trained through its engagement with this environment. I will not discuss Conway et al.'s approach further here, but the evidence of neural encoding that I present below could, as I see it, be understood as providing support for the kind of account they develop.

[6] Macaques are assumed to have very similar visual systems to those of humans, and are widely used in research on the neurophysiology of vision. This is not usually much discussed in the literature, but for a comparison of V4 in macaques and humans, see Roe et al. (2012, 13).

[7] In the literature I discuss, patches are also called "micromaps", which can be somewhat confusing, as patches are not retinotopically structured—that is, they do not preserve the arrangement of a portion of the retinal image, as do many arrays of neurons in the visual system (Li et al. 2014, 210–211). Instead, they earn the name "micromap" (e.g., Chatterjee et al. 2021) through being elements of larger structures that can be understood as isomorphic to—and so maps of—colour space. Patches are also, confusingly, called "columns", especially when described in the context of hypercolumns, of which they are constituent parts (e.g., Ts'o et al. 2009).

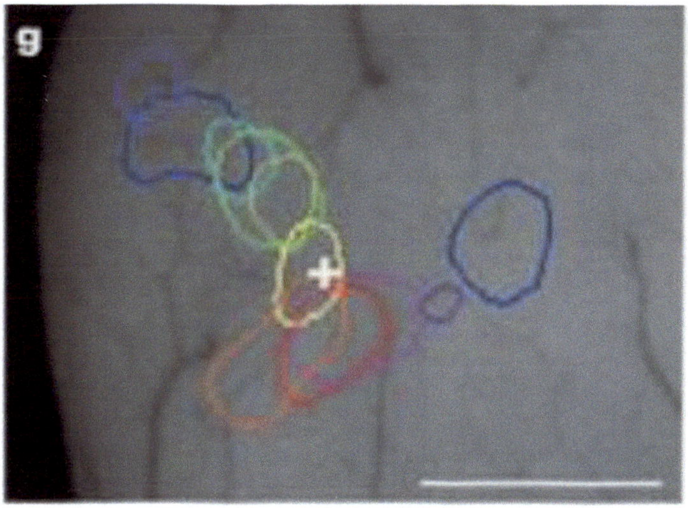

Fig. 5.1 Colour responsive patches in V2, from Xiao et al. (2003), Fig. 1(g)

sensitive to every quality of that kind. The idea that information could be processed in "colour hypercolumns", comprised of layers of colour-sensitive patches, was also not new—Roe and Ts'o correctly suggested that these would be found in regions known as "blobs" in V1 and V2 (Roe and Ts'o 1995, 3713), and Dow (2002) developed a hypothetical account of their structure (although this is now superseded by the empirical evidence I have discussed).

The function of hypercolumns in the visual system is to detect the presence of certain kinds of features in the visual field, and make that information accessible to other perceptual and cognitive processes, whether through direct connections or via the mediation of other processes. Ts'o et al. (2009), Roe and Ts'o (1995), and Chatterjee et al. (2021) have found that the opponent channels feed directly into blobs in V1 and V2.[8] As Chatterjee et al. say, "there is substantial convergence of red/green and blue/yellow inputs to many of these micromap [i.e., patch] cells" (Chatterjee et al. 2021, 6). So, it is reasonable to think that the hypercolumns are encoding information about colour drawn from the opponent channels in a new form, and making it accessible to other perceptual or cognitive processes.

Another group of researchers, Li et al. (2014), appears to have made further relevant discoveries. Where the studies I have looked at so far searched for patches responsive to spectral colours, Li et al. also looked in V4 (in structures called globs, named for their anatomical similarity to blobs) for patches responsive to other colours. Figure 5.2 shows an image from their study that records 37 colour-sensitive patches in a single curling, worm-like form, around 1.5mm in width, that links up with itself,

[8] It was earlier thought that individual blobs were responsive to only red and green or blue and yellow stimuli—that is to say, inputs from a single opponent channel (Ts'o and Gilbert 1988). But this view has since been superseded.

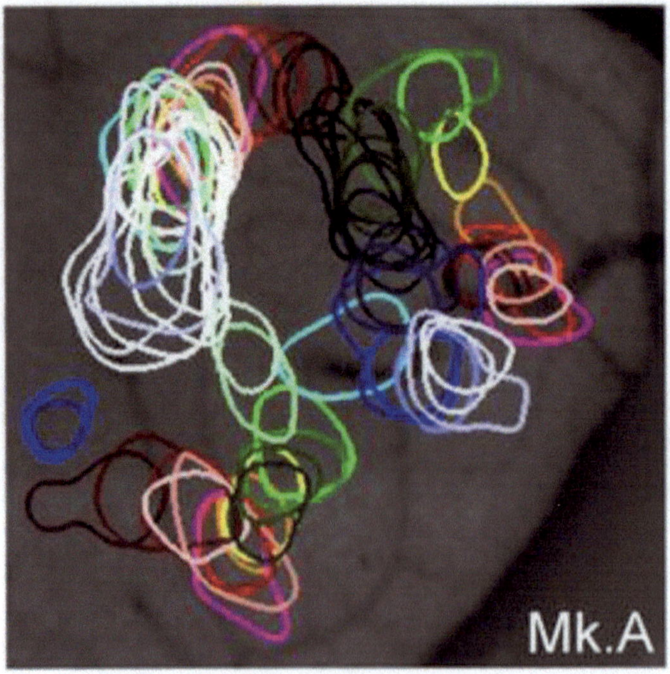

Fig. 5.2 Colour responsive patches in V4. From Li et al. (2014), Fig. 12(a)

and splits, forming new branches. Here the sensitivities of the patches correspond to points not only on the colour wheel, but correspond to tracks over the surface of orthodox colour space (such as Runge's colour sphere or the NCS colour space), the winding columns of patches tracking their way from the chromatic colours through to black and white, and including greys, light and darkened versions of the chromatic colours, and pink and brown. Other images from Li et al. show columns picking somewhat different routes around the surface of orthodox colour space (Li et al. 2014, 215). So, it is apparent that colour processing channels may pass through many such structures in V4, allowing both a colour's hue and intensity to be encoded in this way. Saturation appears to be encoded by these structures too. Li et al. found that variations in saturation, which occupy the interior of orthodox colour spaces, correspond to intensity of activation within corresponding colour patches (Li et al. 2014, 212–213).

One may wonder why the findings I have described about V1, V2 and V4 have not yet been taken up by those who have sought a constitutionalist answer to the question of how elementary hues are encoded in the visual system. These findings are, of course, still relatively new, but there is also another reason why they could be overlooked. If one assumes Hering's account of the elementary colours, then it is understandable to focus on opponent processing in seeking a bearer of information about our colour awareness rather than activity in V1, V2 and V4. From this point of view, it would also make little sense to seek such bearers among the wide range of sensitivities of patches in V1, V2 and V4, since they do not accord with Hering's six elementary colours. However, if we accept a wider range of elementary colours, then it becomes easier to see how structures of this kind could be bearers of colour information.

* * *

Below, I aim to establish three conclusions: (i) that a range of elementary colours wider than Hering's could be encoded in such structures, (ii) that these structures could also encode mixtures of these elementaries, and (iii) that these mixtures could include impossible colours, that is, reddish greens and yellowish blues.

First, some provisos. I do not give an exact account of how colours are encoded—my aim is only to show that it is reasonable to think such encoding occurs in some way using these neural structures. That is also to say that my proposals will naturally raise questions about how the functions I describe could be engineered. It is beyond the scope of this study to answer these, but I hope that it will also be clear that these problems are tractable, even if their solutions are currently unclear. The limitations of the scientific data must also be kept in mind. There might, for instance, be more patches yet to discover in some areas of the visual system. (For example, it would be unsurprising if the hypercolumns in V1 and V2 had a similar complexity to those Li et al. examined in V4). Scanning methods currently in use do not allow connections between the patches, and inputting and outputting from the hypercolumns, to be imaged. Of course, the data I use is from macaques, and so it should be borne in mind that their colour phenomenology and physiology could be somewhat different from our own.

It will also help to clarify my terminology and the commitments that implies. I have been discussing the encoding of information about colour experience, and the neural bearers of this information—patches of neurons, which when activated, indicate the presence of a particular colour feature. These bearers typically—but not necessarily[9]—bear information transmitted by perceptual inputs to the visual system and make this information available to other parts of the visual system and, directly or indirectly, to cognition. Sometimes, I will use the term "representation", as do some of the sources I quote. I use "representation" simply as a synonym for these bearers of information, and I understand those I quote in the same way. Note too that I do not equate these bearers to what philosophers and neuroscientists call "neural correlates of consciousness" (NCCs). David Chalmers proposes that an NCC for a representational state is best understood as: "a minimal neural representational system N such that representation of a content in N is sufficient … for representation of that content

[9] I say that the neural bearers of colour information do not necessarily transmit information from perceptual inputs because there is evidence that the visual system is activated in other contexts as well, including in dreams, hallucinations and visual imagery (e.g., Koenig-Robert and Pearson 2021; Currie and Ravenscroft 2002).

in consciousness" (Chalmers 2000, 31).[10] We have seen that there is a correspondence between the activation of patches and aspects of colour phenomenology. But this does not imply that patches are, or are part of, NCCs. There are two reasons for this. First, the activity of the patches does not correspond with many ordinary colour experiences that we are familiar with, including experiences of constancy, transparency, and illumination and shadow. Second, it is possible that further encoding occurs elsewhere in the brain, and that it is this, rather than the structures I have discussed, that will satisfy conditions for being an NCC for colour experience. V1, V2 and V4 could then be thought of as another input system, that extracts and delivers information to a location where the processing directly associated with conscious awareness and thought occurs.

How then could elementary colours be encoded in hypercolumns? Fig. 5.1 suggests a straightforward approach: that each patch corresponds to an elementary colour. The sensitivities of neurons in the hypercolumn do not gently grade from one colour to another—instead there is a patch many neurons thick made up of neurons responsive to red, then one responsive to orange, then one responsive to yellow, and so on.[11] Looking at Fig. 5.1, we can see that orange and yellow have their own responsive patches, just as Hering's chromatic elementaries do. So, this approach would allow us to acknowledge orange and purple as elementaries alongside red, yellow, green and blue. Figure 5.2 includes patches responsive to black and white, as well as patches that are responsive to brown, and other possible elementary colours, including pink and grey. Here too, it is tempting to consider each patch as corresponding to an elementary colour. Li et al. think along similar lines: "These results [showing patches responsive to pink and brown independently of red] may explain why we perceive pink, brown, and red as three different colors, rather than one color with different lightnesses" (Li et al. 2014, 215). However, the situation is more complex than these results suggest. Figure 5.1 shows patches that are responsive to other colours also: aqua, a yellow-green, and magenta. Perhaps these are elementary colours for us, or for the macaques, but it seems plausible that for us, some of these are not experienced as elementary colours, but as mixtures of elementary colours. Yellow-green is most obviously a mixture for us, as its name suggests. Figure 5.2 raises the same issue more starkly. It includes many patches responsive to colours that are clearly experienced as mixtures: darkened and whitened versions of the chromatic colours,

---

[10] Presumably, removing N from its context in the brain will prevent its functioning as an NCC, so Chalmers also specifies that N exists in the context of "normal brain functioning" (Ibid.).

[11] Conway and Tsao (2009) inserted electrodes directly into the structures they imaged, to gauge the responsiveness of individual neurons. Amidst elements of "noise"—small and large variations in responsiveness—this showed long runs of neurons which have much the same responsiveness, followed by a sudden step shift in responsiveness, where there was a transition to a new patch (Conway and Tsao 2009, Fig. 3). This is consistent with the structure of the best understood hypercolumns, those in V1 that respond to orientation. These are constituted of patches (also called columns) about 100 neurons thick, mostly sensitive to the same stimulus. An abrupt change in sensitivity occurs with the next patch, the cells that make it up are sensitive to a different orientation, about fifteen degrees apart from the orientation that the cells of the first patch are sensitive to. This goes on, with a similar shift in sensitivity occurring with each new patch (Ts'o et al. 2009, 2791–2793).

and various shades of grey. So, while some patches appear to bear information about elementary hues only, other patches "multiplex" this with other information, to use Chauhan et al.'s (2023) term, encoding data about not only elementary hue, but how they combine with other elementary hues, and about their darkness and lightness. Also, as already mentioned, Li et al. found that all patches appear to encode information about colour saturation. Clearly the extraction of this information is no trivial matter, but assuming that the general account I have outlined is correct, there must indeed be channels through which the visual and cognitive systems extract and further process this information. Moreover, any channel that monitors these hypercolumns for a particular elementary colour—say redness—might not only monitor one kind of patch. Rather, it could base its response on monitoring a number of patches, spanning not only red, but all the patches responsive to reddish colours such as a darker red and light red. The same goes for other elementary colours. A channel that monitors for purple, would presumably be responsive in the same way to "pure" purple patches and patches that correspond to dark and light purple. And a channel that monitors for orange will be responsive in the same way to "pure" orange patches, plus those that correspond to light orange. It would, perhaps, not monitor all the way along the parts of the hypercolumn that leads from orange to black, for a channel that monitors for brown would instead be responsive to these patches. In any case, all I wish to claim is that information that describes important parts of our familiar colour experience can be encoded in these structures and extracted by other parts of the visual system. I do not propose that this is the optimal way of understanding these processes; it is only meant to show that they are possible. That is enough to establish the conclusion I seek here: that the resources exist in the visual system to encode information about the broader range of elementary colours than Hering proposes.

How then can hypercolumns encode the mixtures of elementary colours that make up most of colour experience? A hypercolumn can encode red by activating a red sensitive patch, and on the account I have given, triggering the appropriate monitoring channel, it can encode orange in a similar way. But how does it encode the subtle gradations between orange and red, or the subtle gradations between, say a light and dark green? Li et al. (2014) are aware of this problem: "the physical color space …[is] vastly larger than the number of neurons in a small cortical map [i.e., colour hypercolumn] to encode it", making it "implausible that neural activity in a single small cortical area would represent a single color unambiguously" (Li et al. 2014, 216). A hypercolumn does not in itself have the resources to encode gradations of hue, or gradations of lightness and darkness in colours.

How could such information be encoded? Li et al.'s response is that a "combinatorial and distributed color representation is the likely solution to the problem". (Ibid.) They observe that a number of patches within a hypercolumn typically respond to a single stimulus, So, "any single hue point in the visual field is represented by at least one, but typically more, patches of activity; and … different hues are represented by different multipatch patterns in the hue map [i.e., hypercolumn]" (Li et al. 2014, 210–211). Thus, "the limited representational color space of the …[hypercolumns] is likely massively expanded" (Op. cit. 216). This will also be the case for the many hypercolumns through channels of visual processing in V1, V2 and V4 responsive to stimuli in the same point in the visual field. We can add that a condition of representing a mixture will presumably be that the relevant patches must be activated within a channel corresponding to a particular area in the visual field. Thus, a colour representation for a particular area in the visual field is presumably distributed over many different patches in many different hypercolumns through such a channel, and its colour content is a combination of the different contents borne by these patches.

That provides an account of the encoding of colour mixtures. A stimulus may consistently excite the same kind of hue patches—say red-sensitive patches—unequivocally encoding that colour. But more often a stimulus will excite different kinds of patches, corresponding to the mixture of elementary colours that one perceives. For example, A stimulus might excite hue patches with similar sensitivities, say, some red and others orange, in which case, a red–orange colour is encoded and perceived. Here there will be questions about how degrees of mixture are coded—for example, all the different red–orange tints between red and orange. I will not speculate on how this could occur, except to say that such encoding is possible, since there are various ways that this information could be encoded the visual system (for example, patches could be activated to a greater or lesser degree, or it could be that a greater or lesser proportion of red- or orange-sensitive patches are be activated, etc.).[12]

Relatedly, we will also want to know how any such distributed contents can be integrated into a single representation. Anne Treisman identified this problem in its general form, observing that the visual system processes visual information about the features of an object—such as shape, colour, motion, size and distance—in different locations (Treisman and Gelade 1980; Treisman 1986, 1998). This is known as the "binding problem", and it will be apparent that it also arises with my account of the encoding of colour mixtures: how can coding of elementary colours, whether distributed in a single hypercolumn, or through V1, V2 and V4, be "bound" together, so that they are attributed to a single part of the visual field, or the same area on a surface? Although there is currently no generally accepted solution to the binding

[12] I should also describe how these mechanisms could encode Churchland's stygian colours (Churchland 2005). Recall, these are phenomenal mixtures of black with saturated chromatic colours. Like other colour mixtures, these can be encoded by the activation of colour patches in a channel corresponding to a small area in the visual field. So, the stygian orange produced by Fig. 3.6 results from the activation of black patches and orange patches in such a channel. Ordinarily, we do not see stygian colours because there is a physical bar on it: surfaces that are seen as having a chromatic colour must reflect light, but for a surface to appear pitch black, it cannot reflect light. But Fig. 3.6 allows us to see such colours in the following way. After exposure to one or other of the coloured disks, patches corresponding to the complementary colour—yellow or orange—are activated, producing the after-image of that colour. If one fixates on the black surface while the after-image persists, patches corresponding to black are simultaneously activated, so producing the experience of a stygian colour—here, yellow-black or orange-black. There is more to be said about how after-images are produced in the visual system. Their complementarity to stimulus colours indicates that their origins lie in the fatiguing of the photoreceptors. For an account that emphasises the photoreceptors, see Manzotti (2017); for one that sees cortical mechanisms as playing an important role, see Zeki et al. (2017). The fact that after-images persist while fixating on a black surface, as Churchland's stimuli show, would seem to be evidence of the involvement of cortical activity.

problem, a satisfactory solution to it should also give a solution to this instance of it. There are a range of proposed solutions to the binding problem. Treisman and Gelade (1980) developed one possible solution, and others have built on it or developed comparable ideas (van der Velde and de Kamps 2001; Tsotsos et al. 2008; Vul et al. 2020) or developed different solutions (Singer 1999; O'Reilly and Rudy 2001).[13]

We come now to a particular kind of mixture: reddish greens and yellowish blues. Recall that these impossible colours are "impossible" because there is no way to encode them consistent with the opponent process theory. The physiology I describe here, however, allows them to be encoded in much the same way that any other mixture is. Reddish green is encoded when red and green patches are activated simultaneously in a channel corresponding to a single point in the visual field. The same is true for yellowish blue, which requires yellow and blue patches to be similarly activated.

The process by which these impossible colours come to be encoded will, however, be different from that of ordinary colour mixtures. Perception of colours usually relies on input to the hypercolumns from the opponent pathways, but this opponency prevents input that is coded as both red and green or yellow and blue. So, how can impossible colours come to be encoded in the hypercolumns? The answer will lie in understanding how filling-in—which we saw in Chap. 4 is key to producing the perception of impossible colours—can facilitate the encoding of impossible colours in hypercolumns. Here, I think, is a straightforward explanation. Throughout V1, V2 and V4, the channels containing colour hypercolumns are arranged retinotopically: that is, channels that adjoin one another correspond to adjoining points or areas on the retina (Tootell et al. 1998; Roe and Ts'o 1995; Roe et al. 2012).[14] Moreover, patches in adjoining channels that have the same sensitivity will often tend be activated at the same time, causing them to connect more strongly to one another, so that activation of one tends to activate the other, in the absence of other stimuli that would inhibit this. This is a result of the fact that—in Carla Shatz's famous words—"neurons that fire together wire together" (Shatz 1992, 64).[15]

That, in brief, describes a mechanism of the filling-in process: in channels corresponding to an area of the visual field where colour stimuli are limited, stimulation can instead spread "laterally" between channels, so causing the area to be filled-in by the surrounding colours. That allows us to give an account of the production of impossible colours. When an area of the visual field in which stimuli are limited is bordered by different colours, x and y, the signals from the channels corresponding

[13] Cleeremans (2003) includes useful discussions of the problem and its various proposed solutions.

[14] Li et al. also observe that the hypercolumns as they describe them are consistent with the "global retinotopic organization of V4" (Li et al. 2014, 211).

[15] Indeed, there is physiological evidence for strong connections between blobs, the structures in which hypercolumns in V1 and V2 are found. (Livingstone and Hubel 1984) Shatz's phrase summarizes the principle earlier articulated by Donald Hebb: "When an axon of cell A is near enough to excite a cell B and repeatedly or persistently takes part in firing it, some growth process or metabolic change takes place in one or both cells such that A's efficiency, as one of the cells firing B, is increased" (Hebb 1949, 62).

to those borders may propagate laterally through the channels corresponding to the area of limited stimulation, causing the patches sensitive to both *x* and *y* in the hypercolumns in those channels to be activated. So, the corresponding area of the visual field will appear to be coloured with both *x* and *y*. Where *x* and *y* are red and green, the area will appear reddish green, and where *x* and *y* and yellow and blue, the area will appear yellowish blue.

In the past there has been controversy about the mechanisms underlying filling-in. In particular, Dennett (1992) argued that filling-in should not be explained by recourse to the kind of mechanism I have described. While this could be presented as an objection to my account, I take it that Dennett's proposal has since been decisively countered, so here I will only direct the reader to the relevant sources on this count: Churchland and Ramachandran (1994), and Pessoa et al. (1998). I would add that once one is aware of the recent empirical evidence for the multitude of retinotopic maps in the visual system (e.g., Tootell et al. 1998; Roe and Ts'o 1995; Roe et al. 2012), which science was unaware of when Dennett was writing, one can readily understand how these provide ample resources for producing filling-in effects.

To summarize the findings of this chapter: My aim here was to show that there is scope to provide an account of the encoding of colour information consistent with the existence of elementary colours beyond those proposed by Hering, and reddish greens and yellowish blues. We have seen that there is substantial evidence for structures in V1, V2 and V4—colour hypercolumns—into which the opponent processing pathways feed, which can, plausibly, encode this information. As we saw at the start of this chapter, there is a recognition that opponent processes are not bearers of information corresponding to colour experience. Rather, the states that encode this information "must be associated with … higher-level neural activities (Valberg 2001, 1645). I expect that there will be many details of my account that will need revision, as new data comes to light, but I believe I have said enough to make it clear that we should not be constrained by the opponent process theory in describing our colour experience. That is to say, there is no reason to expect that such higher-level neural activities have hard-wired into them Hering's six elementary colours, or that they will enforce Hering's restrictions on colour mixtures. Thus, we should not doubt descriptions of colour experience simply because they are inconsistent with the opponent process theory.

References

Chalmers DJ (2000) What is a neural correlate of consciousness? In: Metzinger T (ed) Neural correlates of consciousness: empirical and conceptual questions, 17–39. MIT Press, Cambridge, Mass

Chatterjee S, Ohki K, Reid RC (2021) Chromatic micromaps in primary visual cortex. Nat Commun 12. https://doi.org/10.1038/s41467-021-22488-3

Chauhan T, Jakovljev I, Thompson LN, Wuerger SM, Martinovic J (2023) Decoding of EEG signals reveals non-uniformities in the neural geometry of colour. Neuroimage 268. https://doi.org/10.1016/j.neuroimage.2023.119884

Churchland PM (2005) Chimerical colors: some phenomenological predictions from cognitive neuroscience. Philos Psychol 18:527–560
Churchland PS, Ramachandran VS (1994) Filling in: why Dennett is wrong. In: Revonsuo A, Kamppinen M (eds) Consciousness in philosophy and cognitive neuroscience, 65–91. Psychology Press, New York
Cleeremans A (ed) (2003) The unity of consciousness: binding, integration, and dissociation. Oxford University Press, Oxford
Conway BR, Malik-Moraleda S, Gibson E (2023) Color appearance and the end of Hering's opponent-colors theory. Trends Cogn Sci 27:791–804
Conway BR, Tsao DY (2009) Color-tuned neurons are spatially clustered according to color preference within alert macaque posterior inferior temporal cortex. Proc Natl Acad Sci 106:18034–18039
Currie G, Ravenscroft I (2002) Recreative minds: Imagination in philosophy and psychology. Clarendon Press, Oxford
Dennett DC (1992) "Filling in" versus finding out: a ubiquitous confusion in cognitive science. In: Pick HL, van den Broek PW, Knill DC (eds) Cognition: conceptual and methodological issues. American Psychological Association, Washington, DC, pp 33–49
Dow BM (2002) Orientation and color columns in monkey visual cortex. Cereb Cortex 12:1005–1015
Forder L, Bosten J, He X, Franklin A (2017) A neural signature of the unique hues. Sci Rep 7. https://doi.org/10.1038/srep42364
Hebb DO (1949) The organization of behavior: a neuropsychological theory. John Wiley & Sons, New York
Hering E (1964) Outlines of a theory of the light sense (trans: Hurvich LM, Jameson D). Harvard University Press, Cambridge, Mass
Hubel DH, Wiesel TN (1974) Uniformity of monkey striate cortex: a parallel relationship between field size, scatter, and magnification factor. J Comp Neurol 158:295–305
Koenig-Robert R, Pearson J (2021) Why do imagery and perception look and feel so different? Philos Trans R Soc Lond B Biol Sci 376. https://doi.org/10.1098/rstb.2019.0703
Kuehni RG (2012) Unique hues and their stimuli—state of the art. Color Res Appl 39:279–287
Li M, Liu F, Juusola M, Tang S (2014) Perceptual color map in macaque visual area V4. J Neurosci 34:202–217
Liu Y, Li M, Zhang X, Lu Y, Gong H, Yin J, Chen Z, Qian L, Yang Y, Andolina IM, Shipp S, Mcloughlin N, Tang S, Wang W (2020) Hierarchical representation for chromatic processing across macaque V1, V2, and V4. Neuron 108:538–550
Livingstone MS, Hubel DH (1984) Anatomy and physiology of a color system in the primate visual cortex. J Neurosci 4:309–356
Manzotti R (2017) A perception-based model of complementary afterimages. SAGE Open 7. https://doi.org/10.1177/2158244016682478
Matthen M (2021) Unique hues and colour experience. In: Brown DH, Macpherson F (eds) The Routledge handbook of philosophy of colour. Routledge, Abingdon and New York, pp 159–174
Mollon JD (2006) Monge: the Verriest Lecture, Lyon, July 2005. Vis Neurosci 23:297–309
Mollon JD (2009) A neural basis for unique hues? Curr Biol 19:R441–R442
Mollon JD, Jordan G (1997) On the nature of unique hues. In: Dickinson C, Murray I, Carden D (eds) John Dalton's color vision legacy. Taylor & Francis, London, pp 381–392
O'Reilly RC, Rudy JW (2001) Conjunctive representations in learning and memory: principles of cortical and hippocampal function. Psychol Rev 108:311–345
Pessoa L, Thompson E, Noë A (1998) Finding out about filling-in: a guide to perceptual completion for visual science and the philosophy of perception. Behav Brain Sci 21:723–748
Roe AW, Chelazzi L, Connor CE, Conway BR, Fujita I, Gallant JL, Lu H, Vanduffel W (2012) Toward a unified theory of visual area V4. Neuron 74:12–29
Roe AW, Ts'o DY (1995) Visual topography in primate V2: multiple representation across functional stripes. J Neurosci 15:3689–3714

Shatz CJ (1992) The developing brain. Sci Am 267:60–67
Singer W (1999) Neuronal synchrony: a versatile code for the definition of relations? Neuron 24:49–65
Stoughton CM, Conway BR (2008) Neural basis for unique hues. Curr Biol 18:R698–R699
Tootell RBH, Hadjikhani NK, Vanduffel W, Liu AK, Mendola JD, Sereno MI, Dale AM (1998) Functional analysis of primary visual cortex (V1) in humans. Proc Natl Acad Sci USA 95:811–817
Treisman AM (1998) Feature binding, attention and object perception. Philos Trans R Soc Lond B Biol Sci 353:1295–1306
Treisman AM (1986) Features and objects in visual processing. Sci Am 255:114B—125
Treisman AM, Gelade G (1980) A feature-integration theory of attention. Cogn Psychol 12:97–136
Ts'o DY, Zarella M, Burkitt G (2009) Whither the hypercolumn? J Physiol 587:2791–2805
Ts'o DY, Gilbert CD (1988) The organization of chromatic and spatial interactions in the primate striate cortex. J Neurosci 8:1712–1727
Tsotsos JK, Rodriguez-Sánchez AJ, Rothenstein AL, Simine E (2008) The different stages of visual recognition need different attentional binding strategies. Brain Res 1225:119–132
Valberg A (2001) Unique hues: an old problem for a new generation. Vision Res 41:1645–1657
van der Velde F, de Kamps M (2001) From knowing what to knowing where: modelling object-based attention with feedback disinhibition of activation. J Cogn Neurosci 13:479–491
Vul E, Rieth CA, Lew TF, Rich AN (2020) The structure of illusory conjunctions reveals hierarchical binding of multipart objects. Atten Percept Psychophys 82:550–563
Wuerger SM, Atkinson P, Cropper S (2005) The cone inputs to the unique-hue mechanisms. Vision Res 45:3210–3223
Xiao Y, Wang Y, Felleman DJ (2003) A spatially organized representation of colour in macaque cortical area V2. Nature 421:535–539
Zeki S (1980) The representation of colours in the cerebral cortex. Nature 284:412–418
Zeki S, Cheadle S, Pepper J, Mylonas D (2017) The constancy of colored after-images. Front Hum Neurosci 11:1–8

Chapter 6
Some Notes on Colour Space

Abstract This chapter draws together the account of colour experience developed over Chaps. 4 and 5, and describes how this alters and enriches existing conceptions of colour space. It argues that although this yields a more complex conception of colour space than we are used to, it should nevertheless be accepted, since its complexity simply reflects that which we find in experience.

Keywords Colour experience · Colour spaces · Unique colours · Impossible colours · Colour temperature

What will the colour space associated with my phenomenology be like? Two factors play a large role in this: the number of elementary colours, and the ways in which they can, and cannot, be mixed. Normal viewers, I have proposed, see at least nine elementary colours, and may see as many as eleven or twelve of them. I have argued that the restrictions on phenomenal colour mixture proposed by Hering do not hold, but this is not to rule out every possible combination of elementary colours from our experience. One may still wonder, for instance, how many elementary colours can be experienced in combination. For example, it seems unlikely that one could ever experience a colour that is a mixture of all these elementaries. But, so far as I can see, this is not due to any underlying principle, such as Hering's opponency. Rather, if there was such a restriction, it would seem to be for more practical reasons. Finding a stimulus that would reliably stimulate the structures described in the previous chapter so that all the different kinds of patches in a channel are activated might well be difficult, or even impossible. However, at a theoretical level—that is, considering the neurophysiology of phenomenal colour mixture that I have sketched—there are no restrictions on colour mixture at all. Any combination of patches may, in principle at least, be activated in a single channel. Because of their distributed character, both within individual hypercolumns and within channels corresponding to parts of the visual field, patches can activate independently: the activity or inactivity of any one patch does not require the activity or inactivity of any other patch. So, the account I have given places no bar in principle on the combinations in which patches may

M. Newall, *Impossible Colours: Colour Experience, Vision Science, and the Philosophy of Perception*, SpringerBriefs in Philosophy,
https://doi.org/10.1007/978-3-032-15137-7_6

be stimulated, and allows, at this level of encoding, that any elementary colour may appear mixed with any other elementary colour.

The number of elementaries, together with any practical restrictions on mixing, will shape colour space, and in potentially complex ways. I will not dwell on details around the number of elementaries and restrictions on their mixture any further here. Instead, I focus on the more generic picture of colour space that my account suggests. The generic picture, as I see it, is this: there may be any number, *n*, of elementary colours, and they may be combined with each other in any way. That gives us a simple and elegant way of thinking of colour space: it will be *n*-dimensional, with each dimension corresponding to an axis on which are located degrees of presence and absence of each elementary colour. (If in practice it turns out that there are some restrictions on the mixtures of elementary colours, the space will be reduced accordingly—but as I say, I put that aside here). This model of colour space will accommodate all the elementary colours we can see. It also allows these to be combined in any way, so providing space for all mixtures, including impossible colours. Note too that there will be separate spaces for brown (which has a dimension assigned to it), and the stygian yellows and oranges (which mix yellow and black, and orange and black, respectively), allowing them to be encoded as distinct colours.

Colour spaces also typically encode other features of colour, and below I examine how the generic colour space I have described treats some of these. I consider (i) colour similarity, (ii) the differing brightness of hues in their saturated states, and (iii) colour "temperature".

I appealed to colour similarity as an organising principle in constructing the colour wheel (in Chap. 2). It will be apparent that more complete colour spaces can be constructed using the same approach. Take all the colours that Hering thinks one can discern, and arrange them so that those that appear more similar to each other are closer to each other—this will produce an orthodox, three-dimensional colour space, such as Runge's *Farben-Kugel* and the NCS colour space. Equally, take all the colours that on my account we can in principle discern, arrange them by similarity, and it will produce the *n*-dimensional colour space I have described. What does similarity mean here? For Hering, similarity is determined by colour composition: orange is similar to red on account of having a reddish component. The less of its yellowish component that it has, the more similar to red it becomes. The idea is visualized in his colour wheel (Fig. 2.1), and, as we have seen, the ways in which Hering's elementaries can be combined determines the shape of the colour wheel and orthodox colour space. That is to say that two colours' similarity or difference can be understood as depending in part on their sharing or not sharing elementary components.[1] Here, I take that idea over from Hering, applying it to the broader array of elementary colours I have argued that we can see. Now, for Hering, one might wonder if that implies that his elementary colours, since they share no components with each other, are all

[1] The idea of colour similarity can be developed further to provide measures of perceptual similarity. Hardin describes how this approach is developed in the NCS system (Hardin 1993, 116–120). The protrusions and irregular form of the Munsell colour space is also the product of incorporating such measures of perceptual similarity.

maximally different from one another. The answer is no, for there are features beyond their different elementary character that set them apart from one another. Focusing on the hues, we find that yellow in its most saturated form is the bright, blue in its most saturated form is the darkest, and red and green, in their most saturated forms, are intermediate in tone. So, some elementary hues are similar or different to others with respect to their degree of brightness in their saturated form (Hardin 1993, 129; Hering 1964, 61–65). Another feature of colours that allows us to identify them as similar or different is colour temperature. This is not a feature that drew Hering's attention, but it will be apparent that of Hering's elementary hues, red and yellow are warm colours, and green and blue are cool colours. So, some elementary colours are similar to or different from others with respect to temperature. (Such colour temperature distinctions are cross-cultural (Hardin 2000, Knoblauch et al. 2023)).

What about my broader range of elementary colours? They too share no elementary components, and in that limited sense are also maximally different from one another. It is also clear that they will be similar to and different from one another by virtue of sharing or differing in degree of brightness in their saturated forms, and colour temperature. Perhaps there is an awkward implication here. Compare orange to red and yellow. For both Hering's position and mine these colours can be distinguished by relations of brightness in their saturated forms (yellow is brightest, red darkest, and orange intermediate), and by shared temperature (all are warm). But for Hering, orange is also similar to red and yellow in virtue of having both red and yellow components, while for my account orange is an elementary colour and is entirely different from red and yellow in this respect. I have already argued in Chap. 3 that at a phenomenological level this is just how things are with orange. However, there is a more difficult case to test our intuitions. Orange is maximally different from red and yellow in terms of its components, but it is also maximally different—which is to say different in the same way and to the same degree—to all the other colours I class as elementary, including green and blue. But green and blue are usually thought to be much more different from orange than from red and yellow, which by comparison we think of as relatively similar. Yet my account says that since they share no components, orange is as different to yellow and red as it is to green and blue. Here, I see no alternative but to grasp the thistle. On my account the apparent similarities and differences here cannot be accounted for in terms of sharing or not sharing components, since they are all the same in this respect, sharing no components. It must be by other kinds of similarities and differences, such as similarities and differences in the degrees of brightness that colours have in their saturated form, and similarities and differences in colour temperature, that explain orange's apparent similarity to red and yellow, and its apparent difference from blue and green.[2]

I now turn to how my account of colour space treats the degree of brightness of chromatic elementary hues. We will see that it can represent this feature of colour,

---

[2] Other features that could play a further role in producing the strong sense of difference between orange, and green and blue, are the fact that, when placed side-by-side they strengthen each other through contrast effects, and produce a distinctive 'vibrating' or 'shimmering' visual effect as a result of chromatic aberration.

although it does not necessarily do so. It will be apparent that some colour spaces do not recognise the different brightness and darknesses of saturated hues—Runge's *Farben-Kugel* and the NCS colour space do not. But some do, such as Munsell colour space (Birren 1969, 73–74).[3] Munsell colour space recognises this by placing all colours of the same tone on the same layer. Saturated yellows, which are lighter, are thus positioned higher in the colour space, and saturated blues, being darker, are positioned lower. This contributes to Munsell colour space's lopsided appearance: saturated yellow occupies a protrusion at the top, while saturated blue occupies a protrusion at the bottom.[4] Not all colour spaces encode this feature of colour—the Munsell space stands apart from many others in this respect, such as NCS colour space. My colour space can also be arranged to accommodate such features. Saturated yellow can be placed so that it is on the same layer as other brightly toned colours, creating a protrusion akin to that in the Munsell space, but in a higher dimensional space, and saturated blue can be placed on the same layer as other similarly dark colours, creating another higher-dimensional protrusion. Like blue, brown will have its own protrusion in the darker part of the colour space, and Churchland's stygian colours will also occupy similar protrusions (as will his "self-luminous" and "hyperbolic" colours (Churchland 2005)).

We speak naturally of warm and cool colours: blue and green are said to be the cool colours; orange, yellow and red are often said to be warm. The distinction seems to hold across cultures[5] (Hardin 2000; Knoblauch et al. 2023). Temperature is a feature of colour that orthodox colour spaces encode. One half of the colour wheel contains the warm colours, and the other half the cool colours. Orthodox colour spaces are similarly divided into a warm part, and a cool part. It will be apparent, however, that my colour space does not group colours by temperature. This is because I allow that elementary colours can be mixed in more ways than allowed by Hering. In particular, the warm and cool colours that Hering thought immiscible—reds and greens, and yellows and blues—can be mixed. So, while my colour space does not elegantly divide into a warm half and cool half, this simply reflects the phenomenological reality of impossible colours.

This raises another question—one that may seem a little whimsical. What temperature, if any, are reddish green and yellowish blue? Do their temperature components cancel each other out? Or are they at once warm and cool? If the latter, one may wonder, is that apparent contradiction a concern? It could be, if one required colour

---

[3] The Munsell system, named for its originator, Albert H. Munsell, calls purple, along with Hering's elementary hues, a "principal" colour, arraying these five colours at equal distances around the vertical axis of the Munsell colour space (Birren 1969, 73). However, I can find no indication beyond this that Munsell himself thought purple elementary in any sense.

[4] These protrusions are also due to Munsell incorporating a measure of perceptual similarity which recognises the high (but differing) degrees that the saturated colours are perceived as varying from a chromatic greys of the same brightness.

[5] There is little work on the basis of colour temperature. Exceptions are Hardin (2000) and Knoblauch et al. (2023), both of which draw on an unpublished study by Elan Katra and B. R. Wooten that proposes a physiological basis in opponent processing. In the following footnote, I suggest that colour temperature could be understood as comparable to synaesthetic experience.

temperature to be similar to the *physical* property of temperature. The physical properties of warmth and coolness are mutually exclusive—a substance cannot be both warm and cool at once. If colour temperature were like that, it would mean a colour could not be both warm and cool. But I think it is more likely that colour temperature relates to *experiences* of warmth and coolness. Sometimes an icy wind can feel like it is burning one's face—yet it feels cold at the same time. Thus, *experiences* of temperature can be warm and cool simultaneously. Perhaps colour temperature is in this respect more like the experience of temperature, in that it allows a colour to appear at once warm and cool.[6] Equally, I would not want to rule out the alternative possibility that the temperature components of reddish green and yellowish blue do simply cancel each other out.

Where does my account of a new kind of colour space leave the familiar, orthodox colour spaces? I am not advocating that these colour spaces are redundant. Colour spaces have different purposes and uses. For a psychologist who wishes to provide a colour space in which subjects can locate their colour experiences, the extra dimensions of colour space I have described may in some cases be indispensable. But those who use colour spaces like a palette—designers using graphics programs, a decorator using pantone colours, a painter using a colour wheel—will have no reason to expand their orthodox colour space, even though these colour spaces incompletely map the colours they can see. Why is that? If one uses a colour space like a palette one can only take from it colours that can be reproduced using a physical sample or swatch—a single homogenous stimulus. If one takes examples of all such stimuli and arranges them so that those that appear most similar are closest together, they will form an orthodox, three-dimensional colour space. For image makers, stimuli of this kind remain the elements of design and picture-making. All other visual effects that one can create in one's chosen medium can be produced by manipulating those elements. These include the colours that cannot be fitted within orthodox colour space. So, none of the colours that normal viewers experience need be off-limits to image-makers, but for that to be so, image-makers should be careful not to confuse the limited colours of their "palette" with the broader category of colours that one can experience.

---

[6] Why should colour temperature be primarily related to the experience of physical temperature? A reason for thinking this comes from understanding colour temperature as akin to synaesthetic experience. Synaesthesia links sensory modalities, so that sensory experience in one modality causes sensory experiences in another. Some synaesthetic connections are the product of connections established through associations between sensory experiences early in life. (Grapheme synaesthetes often see letters as having the same colours as those of their childhood alphabet toys (Rich et al. 2005; Witthoft and Winawer 2013)). If colour temperature were established in that way—say, through associations built up in childhood relating hot and cold to different colours (the colours of flames and ice, the colour coding of taps, etc.)—then it would primarily be a link between sensory experiences, rather than between physical temperature and our experience of colour. Bence Nanay suggests that some of our everyday experiences may have a quasi-synaesthetic character similar to this, and I am encouraged in this speculation by his account (Nanay 2018).

References

Birren F (ed) (1969) A grammar of color: a basic treatise on the system of Alfred H. Munsell. Van Nostrand Reinhold, New York
Churchland PM (2005) Chimerical colors: some phenomenological predictions from cognitive neuroscience. Philos Psychol 18:527–560
Hardin CL (1993) Color for philosophers: unweaving the rainbow, expanded. Hackett, Indianapolis
Hardin CL (2000) Red and yellow, green and blue, warm and cool: explaining colour appearance. J Conscious Stud 7:113–122
Hering E (1964) Outlines of a theory of the light sense (trans: Hurvich LM, Jameson D). Harvard University Press, Cambridge, Mass
Knoblauch K, Werner JS, Webster MA (2023) Warm and cool reheated. Color Res Appl 48:814–817
Nanay B (2018) Multimodal mental imagery. Cortex 105:125–134
Rich AN, Bradshaw JL, Mattingley JB (2005) A systematic, large-scale study of synaesthesia: implications for the role of early experience in lexical-colour associations. Cognition 98:53–84
Witthoft N, Winawer J (2013) Learning, memory, and synesthesia. Psychol Sci 24:258–265

Chapter 7
Philosophical Implications

Abstract This chapter explores the implications of my account of colour experience for the philosophy of colour. It looks at the implications for three general types of theories of colour: objectivism, subjectivism and relationalism. Of these, objectivism finds itself under the most pressure from my account, and we will see that pluralist interpretations of objectivism are best equipped to respond to this challenge. Finally, I return to the ecological and constitutionalist accounts of elementary colours, and show that my understanding of colour experience provides a way of reconciling their core insights.

Keywords Colour experience · Theories of colour · Colour objectivism · Colour pluralism · Origins of the unique colours

This final chapter outlines philosophical implications of my account of colour experience. First, I examine the implications for the three general types of contemporary theories of colour: objectivism, subjectivism and relationalism. Objectivism, at present the most popular of these theories, faces the most serious challenges in this context. Second, I look at how my account of colour experience sits with the ecological and constitutionalist understandings of elementary colours.

* * *

Until now, I have mostly been speaking of colours as experiences. Philosophers understand colour experiences in different ways, and in this way give different philosophical theories of colour. *Objectivism* claims that veridical colour experience is an experience of properties that exist beyond our visual systems, in the external world. Colours, on this account, are properly understood as objective properties that inhere in the things we ordinarily attribute them to, and that stimulate the visual system in a way that gives us a direct experience of them (e.g., Hilbert 1987; Allen 2016). *Subjectivism* holds that colour is simply the experience of colour: a kind of mental property, generated by the visual system. These we project onto objects in the external world, where the world beyond the visual system is not in fact coloured at

M. Newall, *Impossible Colours: Colour Experience, Vision Science, and the Philosophy of Perception*, SpringerBriefs in Philosophy,
https://doi.org/10.1007/978-3-032-15137-7_7

all (e.g., Hardin 1993). *Relationalism* aims to preserve attractive aspects of the first two views. For relationalists, the responses of the visual system together with the stimuli in our environment which produce these experiences are essential aspects of colour. Colour, for relationalism, is thus properly understood as a relation between the viewer's response, and the features of the visible world that produces it (e.g., Matthen 2005; Cohen 2009). For Matthen, for instance, the visual system produces distinctive experiences, "colour-looks", and these have developed as responses to certain features in the external physical world, "colour-properties" (Matthen 2005, 252–268).

Each of these positions can be understood as consistent with my results. However, this is hardest for objectivism, for it faces three challenges that the other theories are better equipped to respond to. The first of these challenges comes from impossible colours. Objectivism understands colours as being those properties of a surface that occasion corresponding sensations—red occasions the experience of redness, for instance. Yet there can be no impossibly coloured surfaces—surface colours are limited to those colours found in orthodox colour spaces. Although there are experiences of reddish green, there are no reddish green surfaces. What then can explain our ability to see impossible colours for the objectivist? While red and green correspond to objective surface properties, the phenomenal mixture of these colours, reddish green, need not. This can be justified by understanding reddish green as the non-veridical experience of two objective properties that in reality are only instantiated in surfaces separately. However, this response requires the objectivist to accept an awkward implication. Reddish green has the same two-part, binary, phenomenological character as any other phenomenal mixture of two elementaries, such as yellowish green, say. So, the objectivist must allow that some experiences of colour mixture can indicate a colour mixture in objective reality while others never do.[1]

Subjectivism and relationalism both straightforwardly accommodate the fact that there are no impossibly-coloured surfaces. These positions locate colour all or in part in the response of the visual system. What qualifies a particular physiology to produce an experience of impossible colour? First, consider the case of colour experience generally. If one is both a physicalist and a subjectivist, then the ability to represent the content of colour experiences in the visual cortex is a necessary condition for a visual system to produce those experiences[2] (Hardin 1993, 111).

---

[1] Macpherson (2003) takes a harder line regarding objectivism, arguing that it is wholly incompatible with the perception of impossible colours. As she sees it, objectivism requires that the perception of any binary colour, including an impossible colour, must require the objective existence of that colour, and not merely its components (Macpherson 2003, 59–60).

[2] I refer to physicalism in this paragraph, because some non-physicalist positions, such as dualism, will not be bound by this condition. Cartesian dualism, for instance, does not require neural structures to function as representations of what a subject sees, since the "thinking substance" can presumably manage such work instead. (This is a "representationalist" interpretation of Descartes (Smith 2024)). I should also add that Hardin identifies not merely as a physicalist, but as a materialist (Hardin 1993, 111). There are other positions on color perception that do not require representation, such as direct realism or enactivism, but I put these aside here since it is clear, as we saw in Chap. 4, that there are neural representations—by which I mean neural vehicles—that function as transmitters of colour information within the visual cortex.

If one is a physicalist and a relationalist, a similar condition holds. For Matthen, to see a particular colour is to experience the visual system tagging an external physical property, the "colour-property", with an "internally available designator", the "colour-look" (Matthen 2005, 257, 268). This designator can be understood as akin to Hardin's neural representation, and the existence of the colour hypercolumns and patches from V1 through to V4 show that the vehicles for such representations or designators exist in the visual cortex. Thus, there are structures in the visual cortex that allow impossible colours to be encoded, and so both subjectivism and relationalism have resources to explain how we experience impossible colours in the absence of any impossibly-coloured surfaces.

The second challenge for objectivism arises from a peculiarity about brown (which I discuss in greater detail in Newall 2022). If one takes a colour swatch that is typically identified as brown and holds it in front of a lighter surface, it will maintain its brown appearance. Yet, as many sources observe, if it is placed in front of a darker surface, it may appear orange or yellow (Hardin 1993, 141; Westphal 1982, 419; Feynman et al. 2013, vol. 1, lecture 35, §3). On Hering's view, this poses no problem, since brown simply is a dark orange or yellow. But on my view, the colours experienced here are two *different* colours, one brown, the other an orange or yellow. Moreover, though we might more often see such a colour swatch against lighter surfaces (since brown is often a relatively dark colour), we need not do so, and often do not. So, there does not seem to be any non-ad hoc way to stipulate standard viewing conditions that would allow us to definitively determine the swatch's colour one way or another. My account of brown as an elementary colour thus presents the objectivist with this problem: explaining how a such a swatch—or area on a surface with the same reflectance properties—can have *two* colours. That is a challenge for objectivism, since it is usually understood as attributing at most *one* colour in such conditions. The challenge can, however, be overcome by accepting colour pluralism, a form of objectivism that allows that a plurality of colours can inhere in a surface with a single reflectance value (Kalderon 2007). But here too the objectivist must accept an implication that runs against the intuition often taken to motivate objectivism: that surfaces may not be coloured as they usually appear, but could invisibly instantiate a plurality of unseen colours.

Again, subjectivism and relationalism readily overcome this challenge, since they locate colour either all or in part in the response of the visual system. Thus, they are not put in the position of attributing two different colours to one surface, but can explain the two different colours a surface seems to have as depending on the two different responses of the visual system to that surface.

The third challenge for objectivism comes from thinking not about actual colour experience—as I have up till now—but rather, possible colour experiences. My account suggests that normal viewers could experience eleven or twelve elementary colours in all. Encoding this number of elementary colours is possible because of the physiology of the hypercolumns. But that physiology raises another possibility. The structure of these hypercolumns could accommodate more patches, allowing for elementary colours to be encoded beyond the eleven or twelve I have suggested.

That would presumably require other changes to the visual system. More receptors and opponent channels would presumably be needed to feed in richer colour data, and whatever apparatus within or beyond V4 that processes the information encoded in patches would also need to be enhanced. I don't see any reason why these are not merely engineering problems—which is to say that the enhanced apparatus required in the visual system is possible, it is just an extension or multiplication of the processing resources already present. It may well be that such enhanced visual systems exist. Some animals, including many birds and fish—and, very occasionally, people—have richer colour input than most humans in virtue of having four kinds of photoreceptors rather than the three that humans usually have. (For cases of animals, see Chen et al. 1984; Marshall and Arikawa 2014. For human tetrachromacy, see Jordan et al. 2010; Jordan and Mollon 2019; Macpherson 2021; Newall 2025) Much as colour-blind individuals may distinguish fewer elementary colours than trichromatic subjects (a red–green colour-blind dichromat, for instance, is unable to distinguish red and green), so it may be that tetrachromat animals and people distinguish more elementary colours than do trichromats. They would presumably have a more complex visual functionality along the lines described above, including patches with sensitivities beyond those of trichromats.

Putting aside such possible real-life examples, let me turn to the question of how many patches with distinct sensitivities there could be. In principle, I don't see any way to place a limit on this. Between any two patches with differing wavelength sensitivities, there could always be a third, sensitive to an intermediate wavelength. And assuming that every patch with a distinct sensitivity is responsive to a distinct colour, there could be a correspondingly unlimited number of elementary colours encoded. One might think that there must be other facts that would forbid development in this direction. One such potential constraint could be found in the further neural apparatus needed to input into and make use of the patches' data. But even if such enhancement would at some point become too resource-intensive to be practical, or subject to diminishing returns in terms of its usefulness, I see no a priori reason why it would not remain possible. Just as more patches are possible, there is no reason to doubt that the associated apparatus of input and further processing will also be possible.

Another potential constraint could be found in the colour stimulus—that is, it might be that the external world only provides enough variations in colour stimulus to support the encoding of a particular number of distinctly sensitive patches. But this is not true either: consider the part of the hypercolumn devoted to representing the spectral hues. Just as one can always add another patch to this part of the hypercolumn, so one can always find a distinct wavelength within the visible spectrum to which it can respond, even if it is only incrementally different to wavelengths that adjacent patches are sensitive to.[3] So, there seems no way to place a limit on the number of such patches, and so no way to limit the number of different elementary colours a visual system may possibly encode. This presents the following challenge to philosophical

[3] I argued above that an objectivist account of my results entails colour pluralism—this will further multiply the numbers of colours to which patches may be responsive.

theories of colour. While on Hering's account we only needed to find a place for six elementary colours in our ontology, my account perhaps doubles this number. But where could an unlimited number of elementary colours find a place?

Non-objectivist positions are able to accommodate this profusion of elementary colours in straightforward ways. Subjectivism holds that our visual systems, as they are, are capable of producing all the colours with which we are familiar. If one accepts that, then it is easy to accept that multiplying the resources of the visual system—including photoreceptors, opponent processing channels, patches, and whatever apparatus exists to further process colour information—would correspondingly multiply the number of elementary colours such a visual system can produce. So, if subjectivism can explain how our limited visual systems can produce the elementary colours we are familiar with, it can equally accommodate any number of elementary colours, by explaining it as the product of a correspondingly complex visual system. The situation is similar for relationalism. Recall that relationalism understands colour as a relation involving a particular kind of feature in the external physical world, and a viewer's response to that feature. So, as the viewer's visual system becomes more complex in the way I've sketched out, and its relation with visual stimuli also becomes correspondingly more complex, it will be no surprise that a correspondingly more complex array of colours, including elementary colours is produced.

For the objectivist the situation is, again, more difficult. Objectivism, we have seen, holds that when perceived veridically, colour properties do in fact inhere in the surfaces we see them in. But since visual systems can exist that can process any of an unlimited number of elementary colours, then those colours must all exist. That is also to say, they will be as real as any of the elementary colours familiar to us, inhering in every visible surface regardless of whether anyone has a visual system that allows them to be seen. Thus, the objectivist is committed to an infinity of elementary colours corresponding one-to-one to the infinity of precise wavelength reflectances. Moreover, since most surfaces reflect a wide range of wavelengths, they will instantiate a correspondingly vast, perhaps infinite, number of different elementary colours.

The intuitive, common-sense character of objectivism has always been a major attraction of the view. But the infinity of elementary colours entailed here, like the earlier considerations I raised, contradicts our ordinary intuitions about colour, and so weighs against objectivism.[4] Perhaps a colour pluralist might wish to grasp this thistle too, but once again, that position can only save colour objectivism at the expense of the common-sense intuitions that inspire it.

* * *

[4] A further counterintuitive implication: the infinity of elementary colours implies that many of the colour mixtures that we can experience are in fact misrepresentations. Wherever we see a colour mixture when viewing a monochromatic light source, there is an actual elementary colour, likely impossible for our limited visual systems to detect, that is the true colour of that light source.

I finish by returning to a topic I touched on in Chap. 5: the ecological and constitutionalist accounts of elementary colours. Say that one has decided between objectivism, subjectivism and relationalism. Unless you have already built an account of elementary hues and phenomenal mixture into your philosophical theory of colour, a question about the nature of elementary colours will remain. Mollon puts it this way:

> [W]hat are the unique [i.e., elementary] hues? Are they determined within us, by the organization of our visual system? Or are they ecologically significant, identifying for us particular subsets of spectra in the world? Let us call answers of the former type "constitutional" hypotheses, and answers of the second type, "ecological". The two types of account are not necessarily exclusive, because our visual categories may have evolved to match some feature of the world. (Mollon 2006, 304)

I put Mollon's question slightly differently, in order to clearly separate it from the theories of colour discussed above. What makes certain reflectances, and not others, correspond to elementary colours for us? Is it, as Mollon says, explained by (i) facts about the constitution of visual system, (ii) by the ecological significance of these reflectances, or (iii) a combination of these—that is, a direct influence in the constitution of the visual system, and an indirect influence, in the shaping of that constitution by our visual ecology?

It will be apparent that there is a correspondence between these three types of responses to Mollon's problem, and the three general theories of colour I considered above. However, a theory of colour does not necessarily commit one to the corresponding account of elementary colours, or vice versa. For instance, one could be a colour objectivist or relationalist but still allow that the elementary character of some colours is purely determined by the visual system.[5] That also means that we cannot look to these theories of colour for a solution to Mollon's problem.

My phenomenology is consistent with each of the three answers Mollon suggests. It fits with constitutionalism, since the coding of elementary colours that I described in Chap. 5 as allowing for my phenomenology could be used to give a filled-out theory of how the constitution of the visual system determines which reflectance properties correspond to elementary colours. My account also allows for the ecological view to be correct, but for that to be so, the elementary colours that my phenomenology describes will require an explanation of why each is prominent in our visual environment and salient to our perception. Moreover, the view may also need to be adapted to incorporate the influences that culture—another aspect of our ecology (albeit not always visual)—may have on determining our elementary colours.[6] Equally, my account is consistent with a conciliatory view, and I think that this provides the

[5] The combination of objectivism and constitutionalism would allow that animals, which can have very different visual systems to ours, may experience the same colours as us (colours being objective features of the world) while experiencing a different selection of those colours as elementary. There is a complexity here in that we should avoid a situation in which one colour is elementary for one species, and (precisely the same colour) is a mixture for another.

[6] Here, recall that different language groups might experience somewhat different elementary colours.

fullest answer to Mollon's question.[7] On this way of thinking, our visual ecology, through a combination of evolution and the conditioning of our individual visual systems, largely shapes our visual systems' sensitivities. So, it is no surprise to find that our visual systems have developed a special responsiveness to distinctive and salient features of the world. In this way, the colours that are for us elementary are determined directly by our visual systems, and indirectly by the visual ecology in which we live, as well as the visual ecologies in which our distant ancestors, human and non-human, evolved.

References

Allen K (2016) A naïve realist theory of color. Oxford University Press, Oxford

Chen DM, Collins JS, Goldsmith TH (1984) The ultraviolet receptor of bird retinas. Science 225:337–340

Cohen J (2009) The red and the real: an essay on color ontology. Oxford University Press, New York

Conway BR, Malik-Moraleda S, Gibson E (2023) Color appearance and the end of Hering's opponent-colors theory. Trends Cogn Sci 27:791–804

Feynman R, Leighton R, Sands M (2013) The Feynman lectures. Basic Books, New York

Hardin CL (1993) Color for philosophers: unweaving the rainbow, expanded. Hackett, Indianapolis

Hilbert DR (1987) Color and color perception: a study in anthropocentric realism. CSLI Publications, Stanford, CA

Jordan G, Deeb SS, Bosten JM, Mollon JD (2010) The dimensionality of color vision in carriers of anomalous trichromacy. J Vis 10:1–19

Jordan G, Mollon JD (2019) Tetrachromacy: the mysterious case of extra-ordinary color vision. Curr Opin Behav Sci 30:130–134

Kalderon ME (2007) Color pluralism. Philos Rev 116:563–601

Macpherson F (2021) Novel colour experiences and their implications. In: Brown DH, Macpherson F (eds) The Routledge handbook of philosophy of colour. Routledge, Abingdon, pp 175–209

Macpherson F (2003) Novel colours and the content of experience. Pac Philos Q 84:43–66

Marshall J, Arikawa K (2014) Unconventional colour vision. Curr Biol 24:R1150–R1154

Matthen M (2005) Seeing, doing, and knowing: a philosophical theory of sense perception. Oxford University Press, Oxford

Mollon JD (2006) Monge: the verriest lecture, Lyon, July 2005. Vis Neurosci 23:297–309

Newall M (2022) A study in brown. Synthese 200. https://doi.org/10.1007/s11229-022-03956-9

Newall M (2025) What is it like to be a tetrachromat? Philos Mind Sci 6. https://doi.org/10.33735/phimisci.2025.11645

Smith K (2024) Descartes' theory of ideas. In: Zalta EN, Nodelman U (eds) Stanford encyclopedia of philosophy (Spring 2024 edn.). https://plato.stanford.edu/entries/descartes-ideas/

Westphal J (1982) Brown. Inq: Interdiscip J 25:417–433

[7] The account of the colour coding in the cortex given by Conway et al. (2023)—which, as I have mentioned, is broadly consistent with mine—also lends itself to such a conciliatory approach.

References

MIX
Papier aus verantwortungsvollen Quellen
Paper from responsible sources
FSC® C105338

If you have any concerns about our products,
you can contact us on
ProductSafety@springernature.com

In case Publisher is established outside the EU,
the EU authorized representative is:
Springer Nature Customer Service Center GmbH
Europaplatz 3, 69115 Heidelberg, Germany

Printed by Libri Plureos GmbH
in Hamburg, Germany